American Institute of Chemical Engineers

Topical Conference on Green Chemical Engineering 2004

AA003076

Held at the 2004 AIChE Spring National Meeting

April 25-29, 2004
New Orleans, Louisiana, USA

Printed from e-media with permission by:

Curran Associates, Inc.
57 Morehouse Lane
Red Hook, NY 12571
www.proceedings.com

ISBN: 978-1-60423-475-6

Some format issues inherent in the e-media version may also appear in this print version.

American Institute of Chemical Engineers

Topical Conference on Green Chemical Engineering
2004

TABLE OF CONTENTS

65 **A Novel Process of Enzymatic Synthesis Coupled with On-Line Separation for Producing N-lauroyl-6-amino Propionitrile through a Packed Bed Reactor**
Yun Fang, Zhi-guo Xu, Yong-an Shen, Yong-Mei Xia and Xue-Yi Hu

Session 143 - Applications of Green Chemistry and Enginering in Manufacturing and the Federal Government

Chair: Ellyn Beary
Vice Chair: Ken Mulholland

Session 144 - From Bench Top to Commercialization: Incentives and Barriers

Chair: Earl R Beaver
Vice Chair: R. Benson Pair, Jr.

Session 145 - University-Based Sustainability Programs

Chair: Eric Beckman
Vice Chair: Harry Tollerton

TWO-STEP METHANE CONVERSION TO ALKYNES AND DIENES

Michael C.J. Bradford, Mure Te and Digna X. Fuentes*
CeraMem Corporation
12 Clematis Avenue
Waltham, MA 02453
Tel: (781) 899-4495
Fax: (781) 899-6478
** Corresponding Author*

ABSTRACT

A novel, multi-step process that employs tailored and regenerable intermediates has been developed for the selective conversion of natural gas to acetylene. The most advanced intermediates developed thus far exhibit an intrinsic production capacity exceeding 5,000 μmoles of acetylene per gram of intermediate per process cycle.

1. INTRODUCTION

In the first half of the 20th century, many valuable chemical intermediates (such as vinyl acetate) were produced using acetylene,[1] which has been primarily used in recent decades as an intermediate for the production of 1,4-butanediol and vinyl chloride monomer. The demand for acetylene could grow appreciably if technical hurdles related to the commercialization of high-conductivity polyacetylene-based materials are overcome.[2] In addition, there is growing domestic and global demand for acetylene as a fuel for welding applications.

Approximately 68% of U.S. acetylene production capacity is based on natural gas pyrolysis or partial oxidation, 18% is derived from ethylene co-production, and 14% is obtained via calcium carbide hydrolysis.[1,3,4] However, each of these technologies has distinct disadvantages. For example, arc and plasma-based routes to natural gas conversion have intrinsically high electrical costs. Pyrolysis-based processes generate dilute concentrations of acetylene and ethylene in the reactor effluent, thus necessitating energy-intensive and expensive downstream separation steps. Product yields are lower than the theoretical limits because intrinsic chemical kinetics and thermodynamic equilibrium generate a broad product distribution, in part because ethylene and acetylene can rapidly oligomerize to aromatics and decompose to coke at the elevated temperatures employed by these technologies. In addition, calcium carbide-based processes are capital and energy intensive. *Consequently, there is a need for an efficient process that selectively converts natural gas to acetylene at moderate conditions.*

Several research groups since 1991 have demonstrated that C_2+ production from methane can be accomplished at moderate conditions using a two-step process.[5,6,7,8] During the first process step, a supported transition metal is exposed to methane at elevated temperature, during which dissociative methane adsorption and carbon-carbon bond formation both occur. Hydrogen is subsequently introduced in the second process step (typically at the same or a lower temperature than the first step) to hydrogenate the surface carbonaceous deposits, facilitate the coupling of surface species, and stimulate product desorption. Although initial work in the area of two-step methane conversion was focused on the production of C_2+ n-alkanes, research efforts shifted during the mid-1990's toward the production of branched alkanes, cycloalkanes, and olefins. During CeraMem's initial work in this area in 1999,[7,8] several technical hurdles were identified that inhibit commercialization of a two-step process for methane conversion to olefins, including:

1. The specific activity of the metal surface generates less than 5 μmol of olefins per gram of supported metal per two-step process cycle;

2. The selectivity to C_2+ olefins during the second process step is typically less than 70%; and

3. A large quantity of excess hydrogen must be used during the hydrogenation step to recover the C_2+ hydrocarbons from the metal surface.

Thereafter, CeraMem conceived of a new technology to overcome these limitations. In 2001 and 2002, the U.S. DOE Small Business Innovation Research (SBIR) program provided CeraMem with Phase I and Phase II awards, respectively, in order to develop a multiple-step process concept for the conversion of natural gas to alkynes or dienes. CeraMem's current objective is to develop a process for the production of two metric tons per day of acetylene from natural gas (Figure 1).

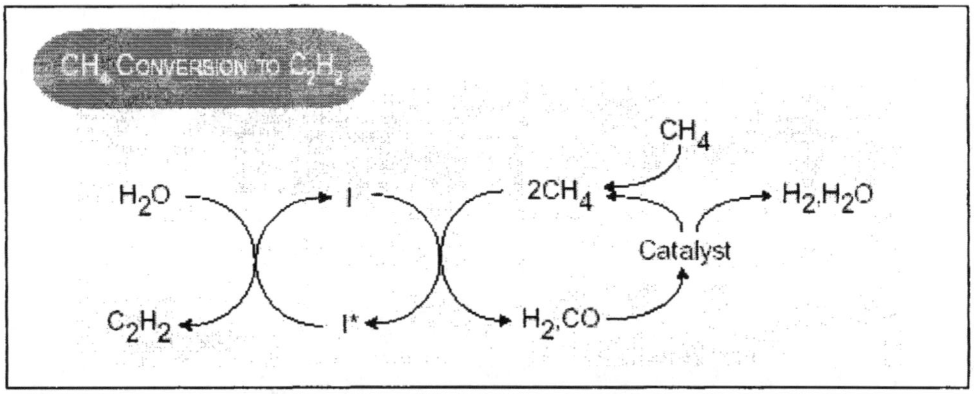

Figure 1. A schematic diagram of CeraMem's process for the production of acetylene from natural gas.

2. RESULTS

The acetylene production for CeraMem's most advanced intermediate (I/I*) is 20x higher than that developed in CeraMem's Phase I program, and over 4000x higher than the observed ethylene production during two-step methane conversion over Ru/Al_2O_3 (Table 1).

Table 1. Observed acetylene production for CeraMem's most advanced intermediate (I/I*) developed in the Phase II program compared with (a) that developed in the Phase I program, and (b) ethylene production observed during two-step methane conversion over Ru/Al_2O_3. [c]

Intermediate (I/I*)	Experimental Results [a]		C_2H_2 Yield / Cycle
	μmol C as $C_2H_2/g_{reagent}$	S (%) [b]	(mol C_2H_2/mol I/I*)
Ru/Al_2O_3 [c]	2.2 [c]	48 [c]	0.0001 [c]
CeraMem "Phase I"	493 ± 88	99.8 ± 0.1	0.01
CeraMem "Phase II"	10,090 ± 338	99.4 ± 0.3	0.43

[a] Reported uncertainties are non-biased standard deviations ($\pm 1\sigma_{n-1}$) and are based on two or more experimental measurements.
[b] Selectivity within the hydrocarbon fraction is reported on a carbon atom basis.
[c] Data refer to the production of ethylene using Ru/Al_2O_3 via two-step CH_4 conversion.[9]

3. SUMMARY

A novel, multi-step process that employs tailored and regenerable intermediates has been developed for the selective conversion of natural gas to acetylene. The most advanced intermediates developed thus far exhibit an intrinsic production capacity exceeding 5,000 μmoles of acetylene per gram of intermediate per process cycle.

ACKNOWLEDGEMENTS

This work was funded by the U.S. Department of Energy under SBIR Grant No. DE-FG02-01ER83161. The authors thank CeraMem Corporation for permission to publish this work.

REFERENCES

[1] L.R.Roberts in J.J.McKetta and W.A.Cunningham (Editors), "Encyclopedia of Chemical Processing and Design." Marcel Dekker, New York, 1976.
[2] *Process Economic Program Report 164: Polyacetylene.* SRI Consulting, 2001. Available at: http://ceh.sric.sri.com/Public/Reports/430.0000/
[3] The Innovation Group: http://www.the-innovation-group.com/welcome.htm
[4] S.Yao, A.Nakayama and E.Suzuki, *Catal.Today*, **71** (2001) 291.
[5] M.Belgued, P.Paréja, A.Amariglio, and H.Amariglio, *Nature*, **352** (1991) 789.
[6] T.Koerts, and R.A. van Santen, *J.Chem.Soc.,Chem.Commun.*, 1281 (1991).
[7] M.C.J.Bradford, *Catal.Lett.*, **66** (2000) 113.
[8] M.C.J.Bradford, *J.Catal.*, **189** (2000) 238.
[9] L.Guczi, G.Stefler, Zs.Koppány, L.Borkó, S.Niwa and F.Mizukami, *Appl.Catal.A:General*, **161** (1997) L29.

NOVEL MEMBRANES AND MEMBRANES PROCESSES FOR INDUSTRIAL SEPARATIONS - PART I

J. Wijmans, K. Amo, R. Baker, R. Blanc, A. Da Costa, R. Daniels, J. He, A. Ariwala and I. Pinnau*
MTR, Inc.
1360 Willow Road
Menlo Park, CA USA, 94025
Tel. (650) 328-2228 x 118
Email: wijmans@mtrinc.com

Abstract

Separations are an indispensable part of petrochemical and refinery operations. Traditional separation processes, such as distillation, pressure swing adsorption and cryogenic condensation, are still dominant. However, the development of novel membranes and membranes processes is increasing the potential of membrane separation processes in these industries.

Over the years past, MTR has executed a number of research and development projects focused on membrane applications in refineries and in petrochemical plants. Most of these projects received financial support from the Office of Industrial Technologies (Program Office of Energy Efficiency and Renewable Energy, U.S. Department of Energy) through its Chemical Industry of the Future initiative and the Small Business Innovation Research Program.

In this presentation we will give an overview of the research and development activities listed above. The applications to be discussed are:

1. Olefin/paraffin separations
2. Hydrogen/hydrocarbon separations
3. Isomer separations

Ultra-Clean Gasoline Technology for the 21st Century

Mitrajit Mukherjee, Exelus Inc., 99 Dorsa Avenue, Livingston, NJ 07039

Introduction

Petroleum refiners worldwide are struggling to comply with environmental mandates that tightly dictate the composition of gasoline and diesel fuel. Reduction in volatile organic compounds, sulfur and nitrogen oxides in automotive tailpipe emissions, and evaporative and refueling emissions have assumed high priority. As a result of these new environmental legislations, refiners all over the world will be required to make significant processing changes. Steps already taken to control vehicle emissions--such as the use of reformulated gasoline--have had a significant effect on air pollution. But stricter regulations and the expected growth in the number of automobiles worldwide in the coming years, especially in developing countries, means that much more needs to be done. Consequently, refiners are re-evaluating their currently available blend-stocks and refinery configurations to determine cost-effective ways to meet market demands and new specifications for cleaner gasoline in light of the recent MTBE ban and the pressing need to lower aromatic and olefin content in gasoline blends. This paper describes the development of a new cleaner-by-design technology to produce a high-octane gasoline component that is environmentally friendly in use. The two main components of this new technology are:

1. A unique solid-acid catalyst that converts light hydrocarbons into an ultra-clean gasoline component and
2. A novel fixed-bed reactor that is designed to enhance the performance of the solid-acid catalyst.

The synergistic integration of the "green" solid acid catalyst with the novel reactor is expected to provide refiners with a cleaner-by-design alternative to produce a high-octane gasoline component – alkylate - to meet the new gasoline requirements.

Importance of alkylate

High-Octane alkylate is the cleanest gasoline blending stream produced in a refinery. Alkylate is an ideal clean fuel component because it is low polluting and has low toxicity. Alkylate has been blended into gasoline for decades to improve octane and thus the antiknock properties of gasoline. Alkylation's importance to refiners continues to grow, and high-octane alkylate has been termed "liquid gold" for reformulated gasoline. It makes up about 13%, or more than 12 million barrels per day, of the current North American fuel. For the refiner striving to meet the stricter motor fuel specifications being mandated on an expanding worldwide basis, investment in alkylation capacity can lead to enhanced refinery flexibility and profitability. Therefore, Alkylate can potentially become one of the most important and cost effective blendstocks of the gasoline pool[1]. Alkylate is currently made using liquid acid catalysts. Refiners typically use either hydrofluoric acid (HF), which can be deadly if spilled, or sulfuric acid (H_2SO_4), which is also potentially harmful and increasingly costly to recycle.

The alkylation reaction combines light olefins with iso-butane in the presence of a strong acid catalyst. It is accepted that alkylation of iso-butane with C_3 –C_5 olefins involves a series of consecutive and simultaneous reactions occurring through carbocation intermediates[2]. The first step is the addition of a proton to the olefin to form a t-butyl cation. The t-butyl cation is then added to an olefin to give the corresponding C_8 carbocation: These C_8 carbo-cations may isomerize via hydride transfer and methyl shifts to form more stable cations. Then the C_8 cations undergo rapid hydride transfer as iso-butane, or other species, regenerates the t-butyl cation to perpetuate the chain sequence.

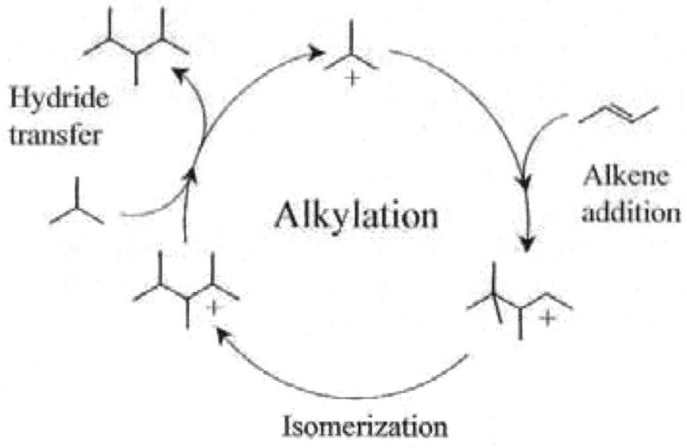

Figure 1. Alkylation reaction mechanism

Unfortunately, these are not the only reactions occurring during alkylation. There are a number of secondary reactions that, in general, tend to reduce the quality of alkylate. Polymerization results from the addition of a second olefin to the C_8^+ cation formed in the primary reaction. The C_{12}^+ can continue to react with an olefin to form a larger cation. The successive addition of olefins to carbocations, or olefin polymerization, is generally believed to be the primary route to catalyst deactivation. The olefin addition reaction is sometimes referred to as a polymerization step, while the hydride-transfer reaction is denoted as the main alkylation reaction. The polymerization reaction results in the formation of “coke”.

Solid acid catalysts have been investigated as alternatives to liquid catalysts for nearly 30 years. Some of these catalysts include $AlCl_3$; platinum compounds; heteropolyacids, such as tungstates; and liquid acids immobilized on silica, polymers, or other solid supports. Natural or artificial zeolites also have been used. Solid catalysts can improve selectivity and reduce production costs, but they tend to deactivate rapidly under alkylation reaction conditions. The problem is buildup of heavier hydrocarbons, known as coke, on the active sites of the catalyst surface. The heavy hydrocarbons also tend to plug the pore structure of solid catalysts, reducing access to acidic sites.

Research on developing a viable solid-acid catalyst to replace HF/H_2SO_4 technologies for iso-butane alkylation has been ongoing since the 1970’s. However, the approach in the past appears to have been one where oil companies have simply screened every known solid acid for its alkylation activity. Indeed, one or more companies have patented all of the known solid acids, yet no viable process has emerged. The problem of rapid deactivation by coke formation has prevented virtually every known solid-acid alkylation catalyst from becoming commercially viable[3].

Solid-Acid Catalyst Issues[4]

Catalyst Activity: A catalytic process is commercially viable if the chemical transformation is achieved with practical limits of space-time yields. Economic necessities determine the lower limit for industrial catalysis – too lond a reactor residence time will make the reactor costs too high. The upper limit is imposed by achievable rates of heat and mass transfer. According to Weisz, the useful space-time yield in the industry lies between 10^{-5}-10^{-6} moles product/cm^3 catalyst/sec for most reactions. Using the lower value of space-time yield, we calculate a liquid hourly space velocity (LHSV) ~ 0.2 1/hr for sulfuric acid catalyzed iso-paraffin alkylation reaction – which is roughly the design value for most commercial reactors. Unfortunately, when one switches from a liquid-acid catalyzed process to one catalyzed by solid-acids, there is a significant decrease in the number of available acid sites. 1 gram of sulfuric acid contains roughly 20×10^{-3} moles of acid sites; 1 gram of solid-acid catalyst (e.g. zeolite or promoted zirconia) on the other hand, contains at most 10^{-3} moles of acid sites of which only 20-30% are strong enough to catalyze the alkylation reaction. In other words, there are 100 times more active acid sites in a gram of liquid-acid compared to a gram of solid-acid catalyst. Once can use a large excess of catalyst to compensate for the lower acid site density for solid acid catalysts – but that would greatly reduce the space-time yields and the process would fall outside the Weisz window, making it economically unattractive. This is one major reason why replacing liquid acids with solid-acid catalysts have proven to be so elusive a goal.

Catalyst Stability: All liquid acids produce conjunct polymers as alkylation by-products. These polymers are called acid-soluble oil (ASO) or red oil. The oil is a complex mixture of unsaturated hydrocarbons having molecular weights in the range of 250-300. They dilute the acid and thus lower the acid strength. In commercial scale operations, a small slip-stream of used acid has to be withdrawn and replenished by fresh acid. In zeolites and other solid-acids, this polymer builds up on acid sites eventually blocking the catalyst pore-mouths and deactivating the catalyst. Pore-mouth plugging typically occurs when the catalyst suffers from severe intra-particle diffusional limitations. The problem of rapid deactivation by pore mouth plugging due to coke formation has prevented virtually every known solid-acid alkylation catalyst from being a commercially viable and environmentally superior alternative to conventional alkylation processes. The diffusion of molecules through catalyst channels and their adsorption on the acid sites is critical to the catalytic action of the solid-acid system. Reactants must be able to access the catalytic sites. Unfortunately, to compensate for the lower amount of acid sites, solid-acid catalysts must have large surface areas (from small pores) to create sufficient amount of active sites. The small catalyst pores (on top of the fast alkylation reaction rate) cause severe intra-particle diffusion limitations for iso-paraffin alkylations leading to very short (on the order of minutes) catalyst lifetimes through severe pore mouth plugging.

As a result of these two critical problems, solid-acid catalysts have failed to be an economically viable replacement to corrosive liquid-acid alkylation processes.

Exelus's approach

Exelus' approach to developing a break-through technology to replace a mature but problematic liquid-acid alkylation process was to leverage its "Engineered Catalyst" (EnCat) technology platform to replace corrosive liquid acids with benign solid-acids. Engineered Catalysts are essentially micro-reactive units, which allow a high level of control of the active site characteristics, transport characteristics and pore architecture at scales ranging from the nano (molecular level) to the macro-scale (reactor level). This higher level of control creates an ideal "reactive environment" at the catalyst surface, which allows the active sites to reach their full potential resulting in performance levels previously considered unattainable on a commercial scale. As a result, the EnCat system is able to make more products more efficiently consuming much less raw material and energy. As catalyst technology and process engineering can now be integrated in a very effective way, the reaction system is able to achieve significantly higher reactor productivity (process intensification) for a variety of chemical and refinery applications. The resulting enhanced catalyst performance leads to lower raw material consumption while greatly reducing energy consumption. This results in lower plant operating costs and a reduction in greenhouse gas emissions.

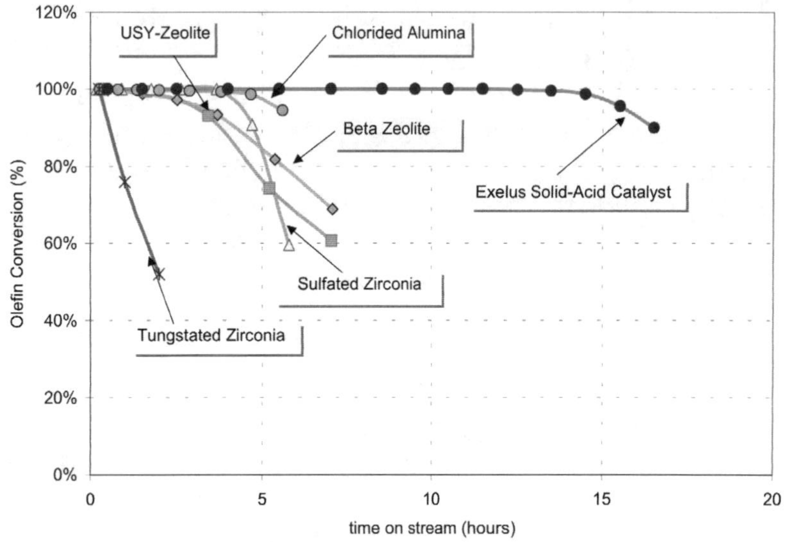

Figure 2. Exelus solid acid catalyst performance compared to conventional catalytic systems

Exelus has used its EnCat technology to develop a "green" catalyst for iso-paraffin alkylation. By clever manipulation of the active material composition (number of sites and acid site strength), researchers at Exelus have developed a unique solid-acid catalytic system that has roughly 400% more active sites than a typical solid-acid catalyst. As a result, the catalyst activity of the solid-acid system is actually higher than a typical liquid acid catalyst, which means that much lower amounts of catalyst are required to catalyze the iso-paraffin alkylation reaction, allowing one to design alkylation reactors with significantly lower volumes (hence cost). By integrating optimized acid-sites with superior mass transport characteristics and a pore architecture that reduces pore-mouth plugging, Exelus has developed a truly revolutionary solid acid catalyst. As shown in figure 2 above, the new catalyst system is easily able to achieve a quantum jump in performance over conventional solid-acid catalysts. All iso-paraffin alkylation tests were run in the same reactor, under identical feed composition (isobutene to olefin ratio and 2-butene as olefin), space velocity, reactor pressures and particle sizes but over a range of temperatures since the modified zirconia and chlorinated alumina catalysts work best at 20-30°C while the zeolite catalysts work best around 70-90°C.

Solid-Acid Catalyst Performance

The robust solid-acid catalyst produces alkylate with a superior octane rating over a wide range of operating temperatures (50-100°C), olefin space velocities (0.1 to 1.0 1/hr) and feed composition (I/O ratios from 6 to 30). Its robustness minimizes the requirement for extensive feed pre-treatment. The technology (catalyst stability/product octane/alkylate yield) has been proven on a bench scale reactor as shown below. Catalyst stability was confirmed by measuring the catalyst performance through multiple cycles of alkylation and regeneration using hydrogen. Figure 3 shows data from a bench-scale evaluation (using a back-mixed reactor) of a single catalyst loading that processed a number of different feedstocks under a variety of process conditions. The data shown are for periods of operation that used similar feed and operating conditions. Feedstocks ranging from pure 2-butylene and isobutylene to butylene mixtures were tested in this run.

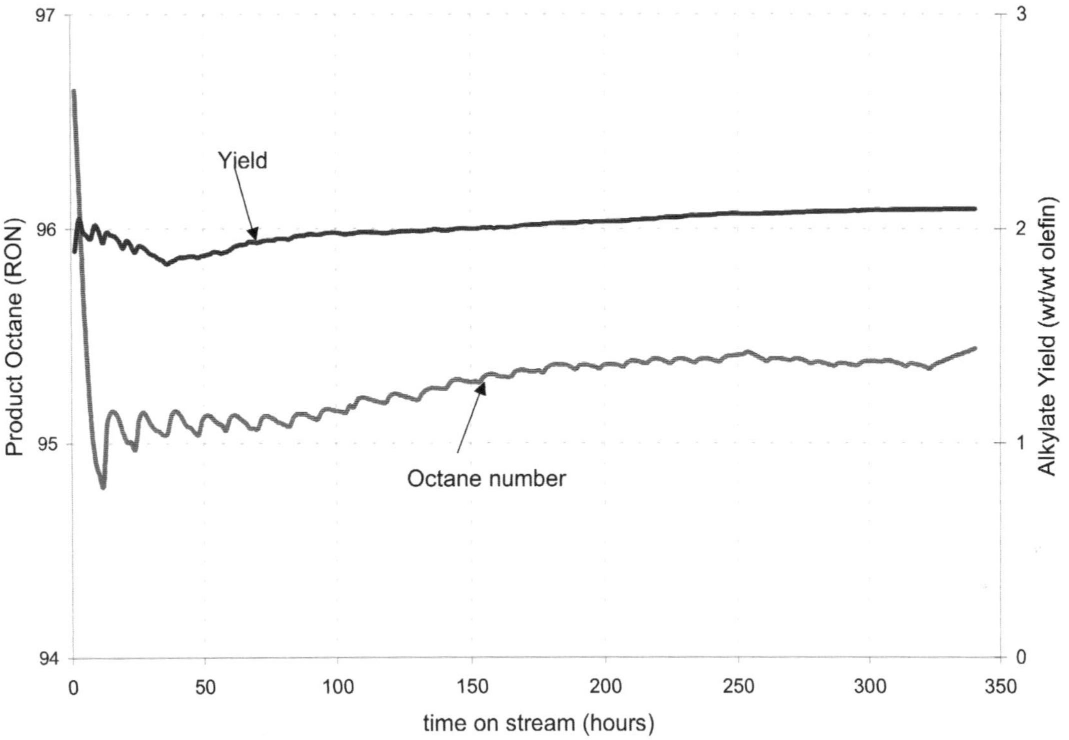

Figure 3. Results of the stability test with using the new solid-acid catalyst

The octane rating (RON) is the same when using 1 or 2-butene but drops by a point when using 30% iso-butylene olefin feed source. Raising the reaction temperature by 30°C reduces the product octane marginally by 1 octane point in contrast to the liquid acid processes, which are penalized severely. The catalyst effectiveness factor also has a large effect on both product slate and catalyst lifetimes. By carefully studying the effect of Thiele Modulus on catalyst life and product octane values, Exelus was able to fine-tune the catalyst diffusional characteristics to further optimize the solid-acid catalyst performance.

Reactor Development

Through an extensive study of the solid-acid alkylation reactions, Exelus has developed a robust kinetic model that is able to predict product octane as well as catalyst lifetimes for a large range of reaction conditions. Estimating intrinsic rate constants was made difficult by the fact that under liquid-phase conditions, the iso-paraffin alkylation reaction is severely diffusion limited[5]. The Exelus model accounts for the effect of intra-particle diffusion and is thus able to accurately predict the rate of catalyst deactivation by through the formation of heavy oligomers. The model has been validated using experimental data, shown in figure 4.

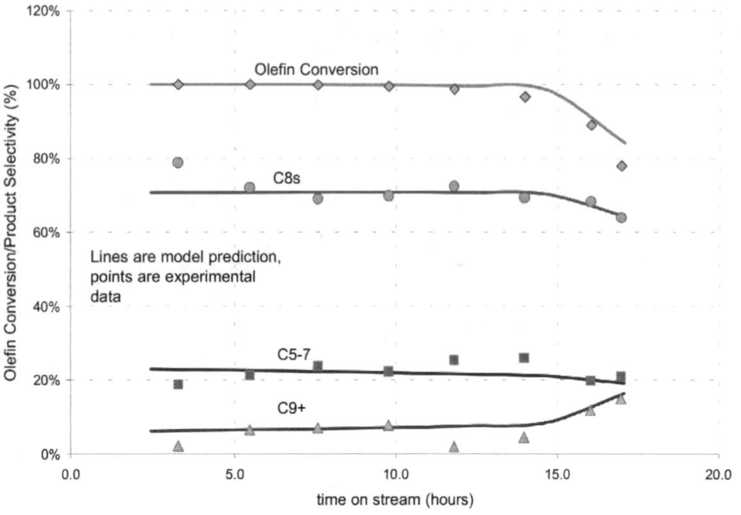

Figure 4. Validation of kinetic model

There are two major competing reactions that occur in iso-paraffin alkylation processes: a) the primary alkylation reaction between an iso-paraffin molecule and an olefin molecule to produce high-octane alkylate and b) the secondary reaction of two or more olefin molecules polymerizing to produce unwanted oligomers which in turn alkylate to form heavy paraffins with low octane rating. Kinetic analysis of these two reactions show that the rate constant for the secondary reaction (k_2) is 2 orders of magnitude higher than the rate constant for the primary reaction (k_1)[6]. To ensure an alkylate selectivity of say 85% (Alkylate/Heavies = 85/15) ~ 5.7, the required iso-paraffin-to-olefin ratio (I/O ratio) in the bulk liquid phase is then ~ 600 mol/mol. The I/O ratio of the alkylation feed stocks ranges from 10 to 15. In conventional liquid-acid based reactors, high-speed mechanical agitators are used to disperse the hydrocarbon feed into the acidic medium. Specially designed jets are used to introduce the olefin feed as small droplets to increase I/O from 8-15 in the feed to 600-1000 in the bulk liquid phase. The only way to achieve the same I/O ratios with solid catalysts is to use a slurry system in a reactor equipped with large stirrers. To ensure an I/O of 600 at the catalyst surface using a fixed-bed reactor, one will need to stage the olefin over 50-60 catalyst beds to meet this target, which is highly impractical in commercial size reactors. While achieving such high I/O ratios is relatively easier in reactor with moving solids (which create the turbulence necessary for olefin dilution) due-diligence on the application of moving solids reactors indicate that such designs will increase cost and complexity of the solid-acid alkylation process The main objective then was to develop a viable reactor design that accomplishes olefin dilution in a fixed-bed reactor while preserving (or enhancing) the product octane value.

Using the kinetic model, Exelus was able to simulate catalyst performance for a large range of reactor configurations. It became clear very quickly that conventional fixed-bed reactor configurations were able to either provide high I/O ratios, or to enhance the product octane number – not deliver both at the same time. In other words, a conventional reactor configuration would not work very effectively for solid-acid alkylation reaction – a new reactor configuration had to be developed.

By applying the kinetic model to novel reactor flow patterns, Exelus has been able to develop a unique alkylation reactor that optimizes both macro-mixing (residence time distribution) and micro-mixing (aggregate mixing) for iso-paraffin alkylation reactions. The macro-mixing profile is optimized to reduce product degradation by minimizing the over-alkylation of iso-octane molecules to heavier (low octane) hydrocarbons.

Table 1. Performance of various reactors for iso-butane alkylation

Reactor	Product Octane (RON)
Plug-Flow Reactor	80 (measured)
Back-Mixed Reactor	95 (measured)
Exelus Reactor	98 (predicted)

The desired micro-mixing characteristics are achieved by using proprietary turbulence enhancing devices called "Dispersion Accelerators" also developed by Exelus. Dispersion Accelerators work to reduce Peclet numbers (i.e. increase liquid dispersion) in alkylation reactors by an order of magnitude over a conventional fixed-bed system by increasing bulk liquid turbulence. This ensures that the reactor iso-butane/olefin (I/O) ratio in bulk liquid is 600 or greater for a feed I/O of 10-15. These unique mixers do not require large recycle pumps to create liquid turbulence, which is another reason why capital costs for the Exelus solid-acid alkylation process are so low. As shown in table 1, the predicted performance of the new fixed-bed alkylation reactor is superior to either a fully back-mixed reactor (CSTR) or a plug-flow reactor. We anticipate that the synergistic combination of the "green catalyst" system with the novel alkylation reactor will be able to produce a high-octane alkylate product in a practical, cost-effective system.

Economic Advantages

The new solid-acid catalyzed alkylation process will have large benefits over conventional liquid-acid alkylation processes. A significant economic advantage of the new solid-acid alkylation process is lower investment costs (about 25% lower) over conventional sulfuric acid processes for the following reasons

1. the fixed-bed reactor design for new process is much simpler than a sulfuric acid contactor reactor,
2. no refrigeration equipment is required and
3. there is no need for brick-lined vessels to separate the acid from the hydrocarbon stream

In addition to lowering investment costs, the new process results in lower operating costs. The sulfuric acid process incurs significant refrigeration costs while ExSact technology operates between 50-100°C and uses low-pressure steam for heating. The Acid-Soluble Oils formed from olefins in the sulfuric acid technology are completely absent in the ExSact process, resulting in lower feedstock consumption. Furthermore, the in-situ hydrogen regeneration of catalyst is significantly less expensive than H_2SO_4 regeneration.

As a result, the new solid-acid catalyst alkylation process has the potential of being an attractive alternative to conventional liquid-acid alkylation technologies.

Conclusions

Several technical challenges have prevented solid-acid catalysts from being economically viable alternatives to conventional liquid-acid catalyzed alkylation processes[7]. By using its proprietary “Engineered Catalysts” platform as a technology enabler, Exelus has developed a unique solid-acid catalyst that overcomes the major technical hurdles facing iso-paraffin alkylation reactions. The new catalyst meets or exceeds (technical) performance requirements for a successful alkylation catalyst without being overly expensive. The catalyst performance has been verified in a bench scale reactor using refinery grade feedstocks and commercially viable WHSVs for over 300 hours on stream.

Exelus has also developed a novel fixed-bed alkylation reactor to enhance the performance of the solid-acid catalyst. This new reactor uses innovative fluid dynamics and an unusual reactor residence time distribution to maximize alkylate production without sacrificing the product octane value. Reactor simulations have confirmed that the novel alkylation reactor affords a much higher octane alkylate compared to conventional reactor configurations.

Acknowledgement

This work was funded by grants from the U.S. Department of Energy and the National Science Foundation.

References

1. Ritter, Stephen K., "Alkylate Rising," C&EN, October 1, 2001, pp. 62-67.

2. Feller, A., Iker Zuazo, Alexander Guzman, Jan O. Barth and Johannes A. Lercher, “Common mechanistic aspects of liquid and solid acid catalyzed alkylation of isobutane with n-butene”, J. Catal. (216), 2003, 313-323

3. Corma, A.; Martinez, A. “Chemistry, catalysts and processes for isoparaffin-olefin alkylation”, Catal. Rev. Sci. Eng. 1993, 35, 483

4. Rao, P.; Vatcha, S. R. “Solid acid alkylation: catalyst and process performance criterion and R&D guidelines”, Oil Gas J. 1996, 9 (Sept), 56-61

5. Simpson M. F, Wei J, Sundaresan S, “Kinetic Analysis of Isobutane/Butene Alkylation over USY Zeolite “Ind. Eng. Chem. Res. 1996; 35 (11):3861-3873

6. de Jong, K. P.; Mesters, C. M. A. M.; Peferoen, D. G. R.; Van brugge, P. T. M.; de Groot, C. Parafin alkylation using zeolite catalysts in a slurry reactor:. Chem. Eng. Sci. 1996, 51 (10), 2053-2060

7. Albright, L. F., “Updating alkylate gasoline technology”, ChemTech, 1998, (6), p40

Green Olefins: Process Development for Energy-Efficient Olefins Recovery

Mike Foral
AIChE Spring National Meeting
April 28, 2004

bp aspen

Outline

> Background: Energy Use in Ethylene Production
> Project Goals and Methodology
> Optimized Advanced Distillation Concept
> Current Status
> Next Steps

bp aspen

Olefins Plant

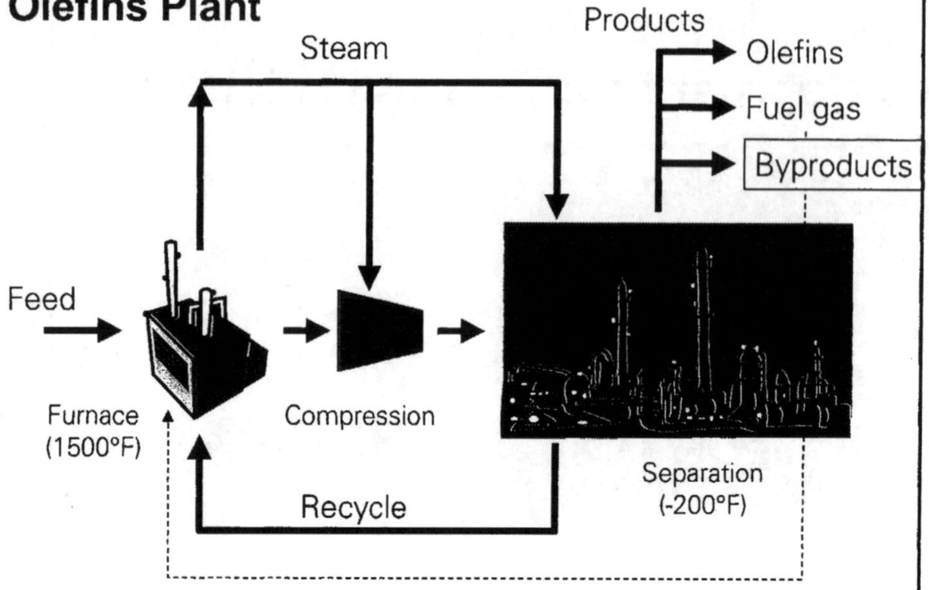

bp

aspen

Energy Requirement for Ethylene Production

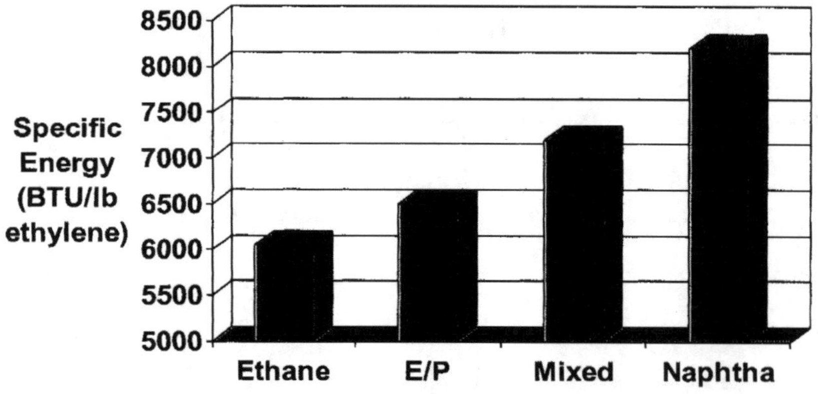

> Greenhouse gas production proportional to energy use

bp

aspen

Project Goals

> Our goals are:

- Produce a new olefins plant separation system that will:
  - Reduce separation energy by 15%
  - Reduce GHG derived from separation section by at least 15%
  - Be at least capital-neutral
  - Utilize only commercially proven technologies
- Develop a new design methodology
  - Start with a blank sheet of paper
  - Combine proven techniques
  - Discriminate between options at a high level

bp aspen

Project Management

> Collaborative project with Aspen Technology

> Project being managed using a stage-gate approach

> Go/No-Go decision after each stage:

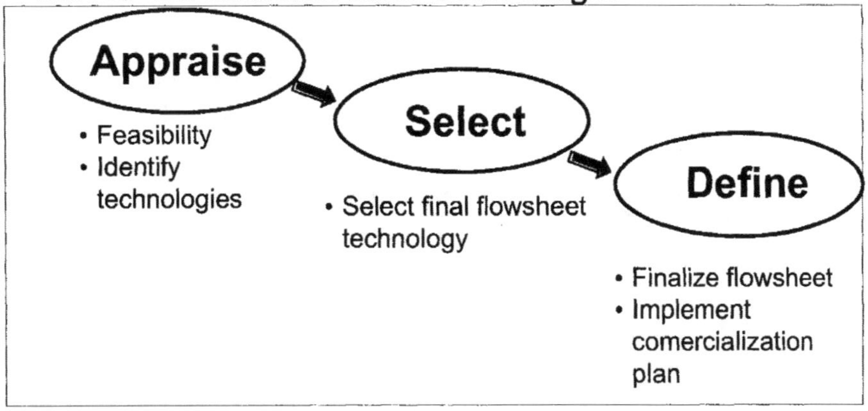

bp aspen

Technology Screening Methodology

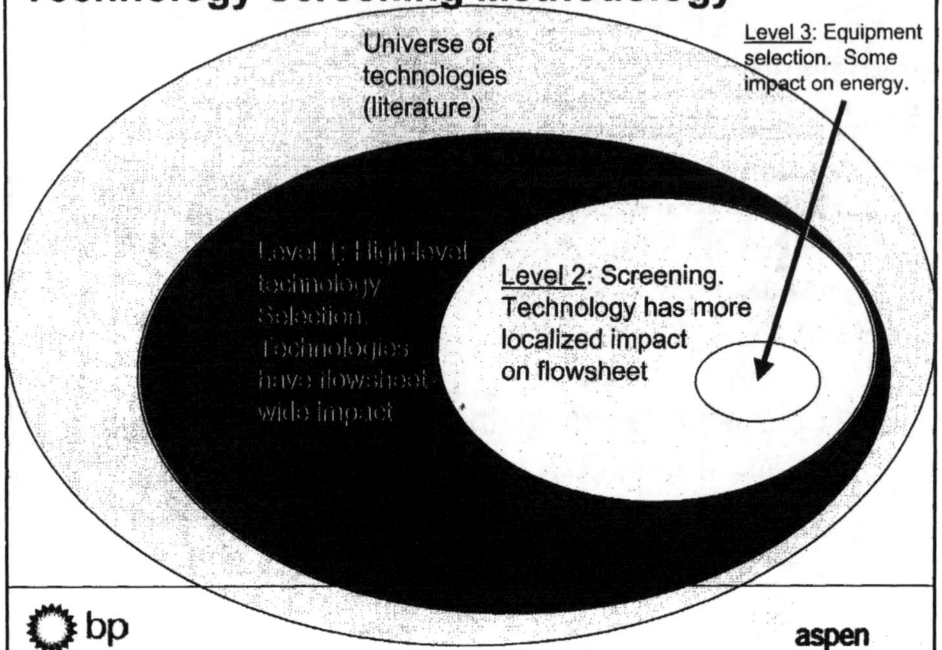

bp

aspen

Level 1 Technologies

> Some of the Level 1 Technologies that were evaluated:
- Absorption
- Distillation (conventional and advanced)
- Hydrogen Membranes
- Mixed Refrigeration Systems

bp

aspen

A Complex Optimization Problem

> Multiple optimization parameters
  - Many column sequencing options: distribution of components and groups of components are possible
  - Sets of columns can be thermally coupled
  - Any column can operate at nearly any pressure
  - Multiple utility (steam and refrigeration) systems are possible

> Problem requires an efficient approach and the right tools.

bp aspen

Green Olefins Advanced Distillation System

> Comprehensive analysis of all feasible column thermal coupling and component distribution options.

> Extensive use of Optimization tools
  - Energy pinch analysis – process, utility integration
  - Column thermal analysis – column operating conditions and design, feed conditioning

bp aspen

Green Olefins Advanced Distillation System

> Careful study of energy/hydrogen production trade-off
  - Determine trade-offs for various flowsheet options
  - Led to novel high-recovery flowsheet design
> Optimized ADS flowsheet, based on energy, flowsheet complexity, capital cost
> Developed simple yet flexible mixed refrigeration system that complements flowsheet design

bp aspen

Conventional Wisdom Says…

> There is a strong trade-off between energy and capital cost.
> Designs providing high hydrogen recovery also have higher ethylene losses
> Designs providing high hydrogen recovery require significantly more energy
> We've been able to break these paradigms by developing a fundamentally more efficient separation scheme…

bp aspen

Current Status

> ADS separation train design with very low energy. 20-25% lower compression power than conventional designs
> Preliminary estimate shows capital savings of more than 10% in columns, compressors, exchangers
> Column and compressor count are similar to conventional designs. No new or unproven technology is employed.

bp aspen

Current Status

> ADS Design flexibility
- Different feedstocks – feasible designs for pure ethane to pure naphtha feeds
- Different furnace technologies (feed conversions)
- Hydrogen production – high hydrogen recovery to product without impacting ethylene loss
- Potential to apply technology to expansions of existing olefins plants. Unique high-capacity debottlenecking options may be possible

bp aspen

There's Still Some Work to be Done:

> Patent protection
> Controllability/operability studies
> Ensure design feasibility of key equipment
> Confirm and implement a commercialization strategy

bp

aspen

Acknowledgements

> We gratefully acknowledge the support of the Department of Energy's Office of Industrial Technology

BP	Aspen Tech
Mike Foral	Rian Reyneke
Brian Bahr	Rik Klavers
James Brocklebank	Gwangsik Jang
Craig Colling	Guang-Chung Lee
Wayne Eng	Iain Sinclair
Jeff Logsdon	
Francisco Lorenzano-Porras	
Christos Papadopoulos	
Graeme Parker	

bp

aspen

Automating the Sustainability Metrics Approach

Dicksen Tanzil, Gang Ma, and Beth Beloff
BRIDGES to Sustainability

As more companies embrace the concepts of sustainable development, there is a need to bring the ideas inherent in "eco-efficiency" and the "triple-bottom line" thinking down to a practical implementation level. Putting this concept into operation requires an understanding of the key indicators of sustainability and how they can be measured to determine if, in fact, progress is being made. These "sustainability metrics" must capture the critical dimensions of sustainability performance needs and translate them into a few measurable quantities that are useful for technical and managerial decision-making, tracking progress, and reporting.

Sustainability metrics have evolved from the conventional EH&S metrics that are compliance-centered, to basic eco-efficiency metrics (producing more value with less), to those that account for potential health and environmental impacts from the facility and along the supply chain (Veleva and Ellenbecker, 2001; Saling et al, 2002; Tanzil et al, 2003). The increase in sophistication and the assessment of potential impacts, however, greatly increases the complexity of metrics calculation, and can be done easily only with the help of automated computational tools.

BRIDGES to Sustainability (BRIDGES), a non-profit organization dedicated to operationalizing sustainable development, has been actively involved in the development of sustainability metrics since 1999, in part supported by funding from US Department of Energy through sub-contract with AIChE's CWRT. Key indicators for the sustainability metrics were identified and heuristics were developed through an industrial workshop and pilots. To further facilitate the use of sustainability metrics and encouraged by industry representatives, BRIDGES has recently automated its approach into a metrics management software tool: BRIDGESworks™ Metrics.

BRIDGESworks™ Metrics identifies key sustainability indicators and offers a variety of metrics for measuring sustainability performance. The tool was designed to offer both convenience and flexibility. It incorporates BRIDGES' set of metrics and their heuristics for calculation, as well as a robust set of impact assessment data for use in identifying pollutant effects. While providing a metrics management starting point, the user has the option of creating other metrics defined by the user. Therefore, the software tool can be easily adapted to simplify calculations for other metric sets.

Following the practice of NRTEE and CWRT, BRIDGESworks™ constructs the sustainability metrics as ratios with environmental impacts in the numerator and a physically- or financially-meaningful representation of output in the denominator. They follow a simple rule of thumb: the small the metric the better. The metrics are currently organized into six basic impact categories: material, energy, and water intensities, solid waste, toxic releases, and pollutant effects. A seventh basic impact category, land use, is currently under development.

The metrics are calculated usually with mass of product(s), sales revenue, and value-added in the denominator. Other representations of outputs, however, can also be used. This includes other units of production (volume, pieces of products, etc.) and functional units such

Table 1. Basic sustainability metrics developed by BRIDGES

Output: *Mass of Product* or *Sales Revenue* or *Value-Added*	**Material Intensity** $\frac{\text{Mass of raw materials} - \text{Mass of products}}{\text{Output}}$ **Water Intensity** $\frac{\text{Volume of fresh water used}}{\text{Output}}$ **Energy Intensity** $\frac{\text{Net energy used as primary fuel equivalent}}{\text{Output}}$ **Solid Waste** $\frac{\text{Total mass of solid waste disposed}}{\text{Output}}$ **Toxic Release** $\frac{\text{Total mass of recognized toxics released}}{\text{Output}}$

as service life. To protect proprietary information, the value-added is defined simply as the difference between sales revenue and the costs of raw materials and utilities. Where confidentiality is not of concern (such as for internal use), one may alternatively use profit margin in place of the value-added denominator. Additionally, value-added can be defined to express the flow of benefits to the community, or the societal value-added. A linkage to Sylvatica's TCAce software to produce more robust financial cost and benefit input data is being developed.

Basic metrics are defined for five of the basic impact categories (all except pollutant effects). These form the simplest set of metrics that can be commonly used across industries and applications. Definitions of the five basic metrics and the pollutant effect metrics are provided in Table 1. Once the boundary for the metrics calculation is defined (e.g. around a process), the metrics can be calculated usually from readily available data.

Complementary metrics can be defined and developed as any other need for decision-support emerges. A common temptation in designing metrics is to take into consideration too many factors. Measurements that combine too many components are actually less versatile and less useful for making comparisons across products and industries. Therefore, it is best to maintain a basic set of simple, widely applicable metrics and to construct complementary metrics to meet the needs of specific industry or application. Yet, the use of complementary metrics remains important to better understand the progress towards sustainability.

Table 2 shows examples of complementary metrics under each category. Some complementary metrics contain impacts not included in the basic metrics, such as the use of packaging materials and energy for transportation, and the amount of rainwater contaminated by the industrial operation. Other complementary metrics help decision makers to better understand the sustainability implications of the resource uses and environmental releases.

Table 2. Examples of complementary metrics under each metric category	
Material ❑ Packaging materials ❑ Non-renewable materials ❑ Toxics in product ❑ Toxics in raw materials Water ❑ Rainwater sent to treatment ❑ Water from endangered ecosystem sources ❑ Water use relative to water availability **Energy** ❑ Energy consumed in transportation ❑ Non-renewable energy	**Solid waste** ❑ Solid waste disposed relative to landfilling capacity **Toxic release** ❑ Toxic release under each TRI category ❑ Human toxicity (carcinogenic) ❑ Human toxicity (non-carcinogenic) ❑ Ecosystem toxicity **Pollutant effects** ❑ Global warming potential ❑ Tropospheric ozone depletion potential ❑ Photochemical ozone creation potential ❑ Air acidification potential ❑ Eutrophication potential

Water usage becomes more critical when it affects an endangered ecosystem, or in a water scarce area. Similarly, solid waste releases can be weighted by the landfill capacity available in the area.

As complementary metrics under toxic release, the potential effects of toxics on human health and the ecosystem can be assessed using available impact assessment methods. Similarly, impact assessment can be applied to calculate the potential pollutant effects on global warming, ozone depletion, photochemical ozone (smog) generation, air acidification, water eutrophication, etc. As mentioned earlier, international consensus has been reached for the calculation of only two of these impacts: global warming and ozone depletion. Other impacts depend on varying degrees on regional and local conditions, although the existing impact assessment methodologies still act a rough guide in weighting the "badness" of different toxics and pollutants.

Data can be manually entered by the user into the BRIDGESworks™ Metrics software system or imported from other sources such as MS Excel. Figure 1 shows the data entry screen for a PVC process. User can enter inflow and outflow data, including:

- rates of usage for raw materials, energy, and water,
- rates of production for both products and salable co-products,
- rates of releases for solid wastes and toxics/pollutants to air and water,

along with their associated costs and revenues. The user can also review the heuristics and assumptions built-in in the software, such as shown in Figure 2 (which shows the default assumptions for electricity, including typical efficiency and fuel mix for electricity generation).

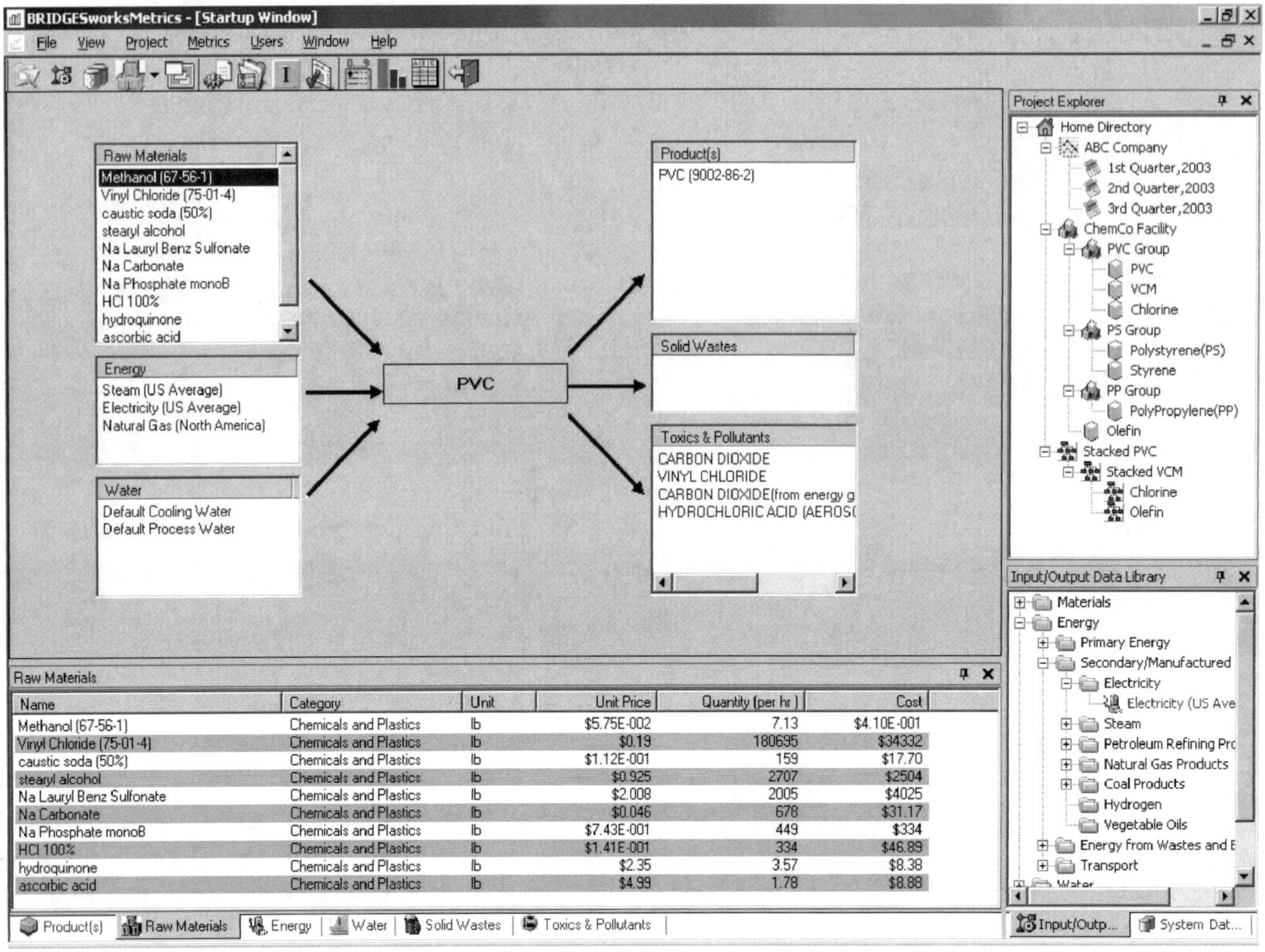

Figure 1. Data input screen for PVC process unit

The software tool can then calculate the basic metrics and complementary metrics, including potential environmental impacts, selected by the user based on relevance to his/her own project. Over thirty complementary metrics and impact assessment methodologies, including EPA's TRACI (Bare et al, 2003), are currently included in the software, with more being developed and incorporated. Users can also add other metrics important to their company.

Once the data are entered and options are selected, one may generate the metrics report. The report can be generated not only for the process unit, but also aggregated for larger business units such as the facility, corporation, etc. In addition, the units of both mass and financial measures can be converted into a variety of denominations. For illustration, let us consider the PVC process in the context of the entire chemical facility (let us call it 'ChemCo', Figure 3). Figure 4 shows the summary reports for the PVC process unit, a 'PVC Group' (collection of process units including the PVC unit), and for the entire 'ChemCo' facility. The metrics can also be plotted and tabulated to allow comparisons between different units, technologies, or time periods (e.g. Figure 5).

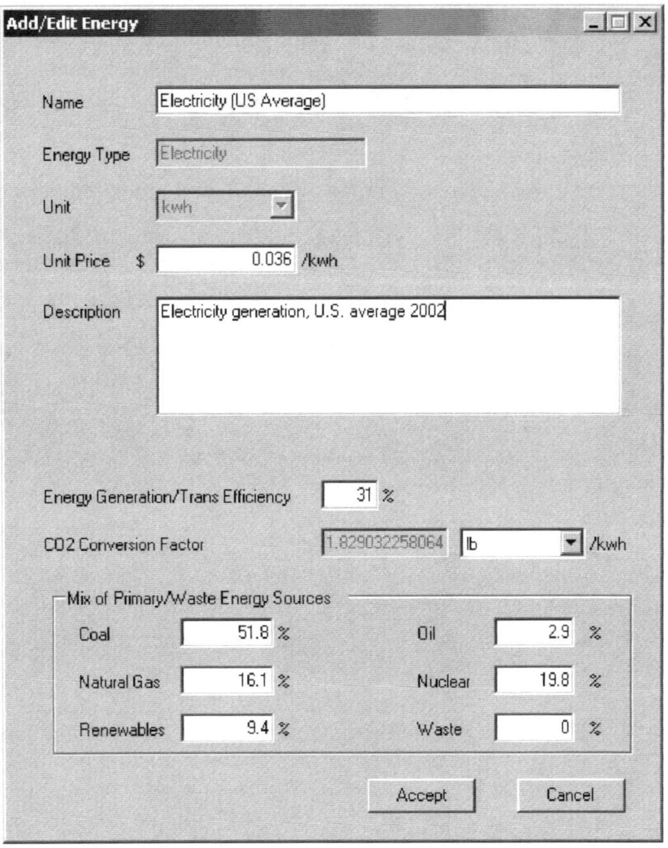

Figure 2. Default assumptions for electricity

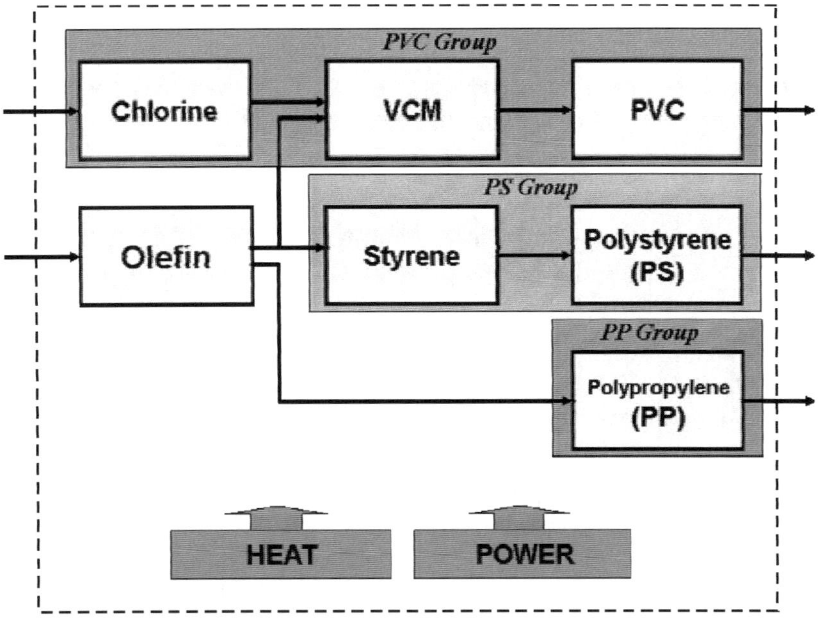

Figure 3. Processes and process groups within 'ChemCo' facility

Figure 4. Metrics summary reports generated for successively larger scale

BRIDGESworks™ Metrics also allow "stacking," i.e. aggregating impacts attributable to a unit of product along the lifecycle of the product system. Stacking can be performed either limited to the internal, or 'gate-to-gate', lifecycle (e.g. processes within 'ChemCo' that contribute to the production of PVC), or extended to the supply chain outside of 'ChemCo' using supplier data or lifecycle assessment databases. Figure 6 illustrates the comparison of the greenhouse-gas emission metric for the entire facility with those attributable to the PVC product. BRIDGES has established a relationship to PRé Consultants, the developer of the lifecycle software SimaPro, and will be developing links to that software in order to facilitate the stacking of data along the product's lifecycle.

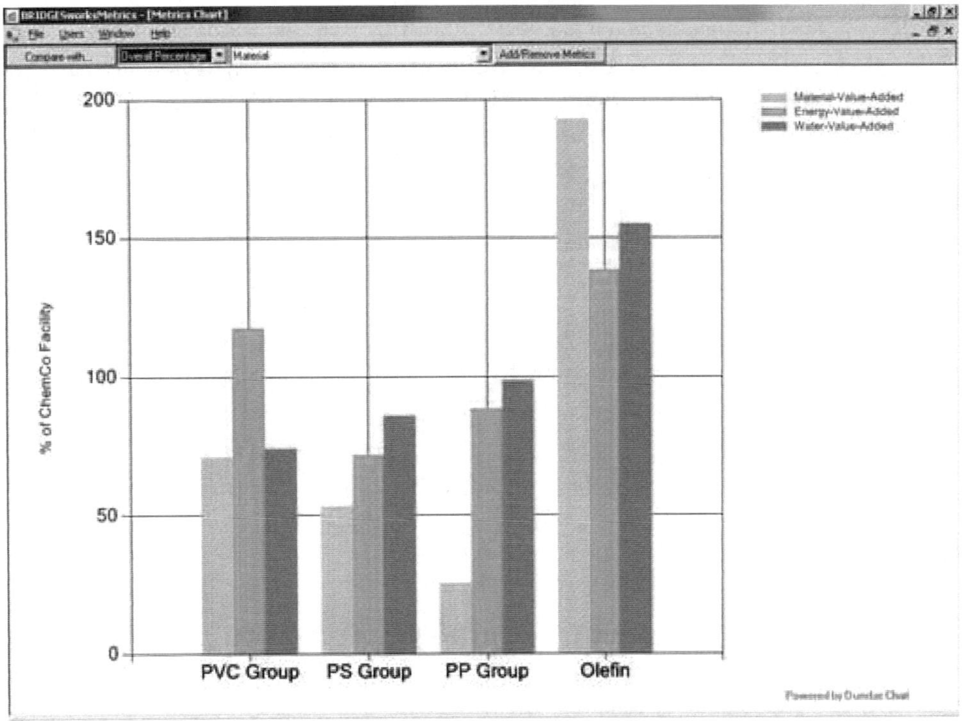

Figure 5. Comparing resource consumption per monetary value-added among process groups within 'ChemCo'

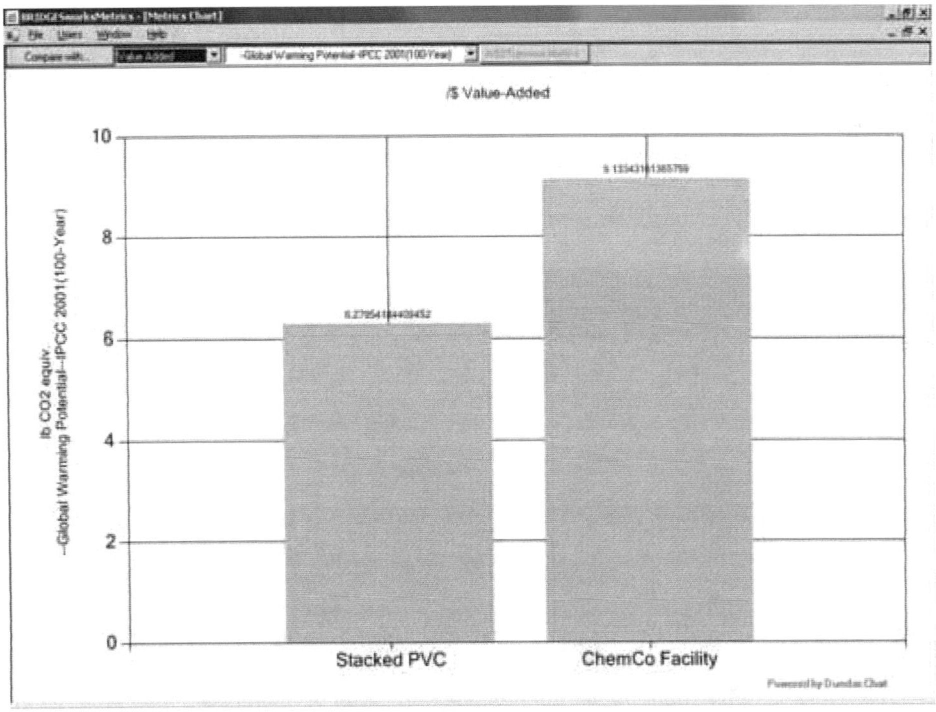

Figure 6. Comparing greenhouse gas emissions attributable to each dollar value-add of PVC to overall 'ChemCo' facility

BRIDGES' sustainability metrics are being further developed for two distinct applications. One is for the educational marketplace, where the metrics management tool is populated with case study information and can be used to educate university students as well as corporate trainees about sustainable development. The other is for corporate users with a variety of potential uses. Ultimately the results will contribute to building the business case for sustainable development.

References:

Bare, J.C., G.A. Norris, D.W. Pennington and T. McKone, "TRACI: The Tool for the Reduction and Assessment of Other Environmental Impacts," *Journal of Industrial Ecology*, **6**(3-4), pp. 49-78 (2003).

Saling, P., A. Kicherer, B. Dittrich-Krämer, R. Wittlinger, W. Zombik, I. Schmidt, W. Schrott, and S. Schmidt, "Eco-Efficiency Analysis by BASF: The Method," *International Journal of Life-Cycle Assessment*, **7**(4), pp. 203-218 (2002).

Tanzil, D., G. Ma, and B.R. Beloff, "Sustainability Metrics," presented at the 11th International Conference of the Greening of Industry Network, San Francisco, October 12-15 (2003). Available at www.bridgestos.org/Publications.htm.

Veleva, V. and M. Ellenbecker, "Indicators of Sustainable Production: Framework and Methodology," *Journal of Cleaner Production*, **9**, pp. 519-549 (2001).

Diesel Emission Reduction Strategies for School Buses: Idle and Mobile Emissions Test Results
A. Toback, J. Hearne, J. Akers, R. P. Hesketh, and A.J. Marchese

Introduction

In response to the Clean Air Act, the US Environmental Protection Agency (USEPA) established National Ambient Air Quality Standards (NAAQS) for various pollutants – known as "criteria" pollutants – that adversely affect human health. The major transportation-related criteria pollutants include Ozone (O_3) and its precursors, volatile organic compounds (VOC) and oxides of nitrogen (NO_X); particulate matter (PM) and carbon monoxide (CO).

Presently, each state is required to submit to EPA a State Implementation Plan or SIP, which is a collection of strategies/ commitments that explain how the State will achieve the air quality standards, set by the Federal Clean Air Act. SIP's are required for a variety of pollutants (Carbon Monoxide, Ozone, Particulates, etc). In New Jersey, the Department of Environmental Protection (NJDEP) is the agency responsible for assembling and submitting the SIP to USEPA. The SIP includes strategies and commitments for stationary (factories, etc) and mobile (auto inspections, etc) sources. The New Jersey Department of Transportation (NJDOT) provides input to the mobile source portion of the SIP. In addition to implementing reduction technologies for mobile emissions from school buses, NJDOT has also identified reduction of school bus emissions as a potential component of future SIP's.

There are approximately 11,000 school buses used each day in the state of New Jersey.[1] These school buses are often idled in the morning prior to their initial bus route and in the afternoon while waiting to pick up the children to take them home. School bus idling occurs for the following reasons:

- to keep the engine and fuel warm in cold weather,
- to provide heat inside the bus in cold weather, and
- to provide power for lighting for safety purposes.

Previous truck idle emission studies have been performed on Heavy Duty Diesel Vehicles (HDDV's). These experiments showed that HDDV trucks produce average emissions of 144 g/hr of NO_X and 8224 g/hr of CO_2.[2] However, the HDDV vehicles have engines that are almost double the size of school bus engines. Therefore, these values cannot be used to accurately estimate the emissions and fuel consumption from school bus idling.

The results from the HDDV truck idling study also show that ambient temperature and humidity have an effect on the formation of NO_X gases.[3] Accordingly, the present study was performed to accurately quantify emissions from school bus idling for various temperature and humidity.

The next phase of testing is the testing of alternative fuels. Historically, chassis dynamometers were used to measure emissions. The chassis dynamometer allows for an indoor use and a controlled climate. In order to more accurately simulate real driving conditions, the school buses will undergo mobile testing.

Alternative fuels such as biodiesel and ultra low sulfur diesel (ULSD) have been shown to reduce some diesel emissions significantly in cars and trucks. In a comprehensive study compiling several separate biodiesel studies, the EPA concluded that common blends of biodiesel containing 20 % by volume biodiesel and 80 % by volume #2 petroleum diesel (referred to as B20) reduced HC emissions by 20 %, PM emissions by 10 %, and CO emissions by 11 %.[4] NO_X emissions increased slightly in some engines (~ 2%) and CO_2 emissions showed little or no difference. The USEPA classifies ULSD has having no more than 15-ppm sulfur content as compared to current U.S. regulations which permit up to 500-ppm sulfur for diesel fuel used for highway transportation. Previous studies show that ULSD by itself reduces HC emissions by 13 %, PM emissions by 13 %, CO emissions by 6 %, and NO_X emissions by 3 %. With the addition of a particulate filter, ULSD can reduce HC and CO emissions by 90 %, PM emissions by 80%, and NO_X emissions by 15 % to 20 %.[4]

NJDOT would like to further explore the potential of alternative fuels and emission reduction strategies through a systematic experimental study of emissions and fuel consumption from a variety of school buses under a controlled test cycle. Accordingly, the NJDOT, in collaboration with Rowan University and Aberdeen Test Center, has initiated an experimental study to quantify the idle and mobile emissions and fuel consumption rates for school buses.

The objectives of this study were as follows:

- quantify fuel consumption rates of school buses at idle and using the Rowan University Composite School Bus Cycle (RUCSBC),
- quantify school bus idle exhaust emissions at various temperatures and humidity,
- determine the effects of ambient temperature and humidity on school bus engine idle emissions and fuel consumption,
- create a new NO_X correction factor[5]
- quantify school bus mobile exhaust emissions using the RUCSBC with 4 fuel options,

Experimental

Three school buses were tested at the Aberdeen Test Center (ATC) in Aberdeen, Maryland. The three buses selected were the statistically most popular buses used in New Jersey.[1] Engine data for each of the school buses utilized in the testing is shown below in Table 1.

Table 1: Engine Data for the School Buses Tested

Chassis	Engine	Rated Power
1998 International	'97 International T-444E	190 at 2300 RPM
1998 International	'97 International DT466	190 at 2300 RPM
1997 Ford	'96 Cummins 5.9L B Series	190 at 2200 RPM

A Sensors Inc. SEMTECH-D mobile emissions analyzer and the Sensors Inc. PM-300 particulate measuring device were used to acquire the above exhaust constituents during each test. The SEMTECH-D is a portable PC-based data acquisition system capable of measuring real time emission levels, ambient temperature, humidity, and engine parameters. SEMTECH-D utilizes a heated sampling line along with the following chemical sensors:[6]

- Flame Ionization Detector (FID) for HC
- Non-Dispersive Ultraviolet (NDUV) for NO_X
- Non-Dispersive Infrared (NDIR) for CO and CO_2
- Electrochemical sensor for O_2.

SEMTECH-D receives engine RPM, oil temperature, and air/fuel ratio from the Engine Control Module (ECM) as part of its vehicle interface. For mobile testing, a Global Positioning System (GPS) and weather probe measure position along with ambient temperature and humidity.

The Sensors PM-300 is a portable particulate measurement system. It operates using a light scattering technology to sort and quantify particles. The air sample is carried into the PM-300 with a heated line and dried prior to analysis. A semiconductor laser emits a beam of light that is detected by a photo diode. Each particle produces a pulse in the beam that is proportional to the particle size. This method allows the PM-300 to sort the particles by size. The particulates are also collected on PTFE filter that can be removed for analysis.[7]

For the idle study, the school buses were tested in an environmental chamber. The environmental chamber is capable of controlling multiple climatic variables, including temperature, humidity, solar radiation, dust, icing, fog, and thermal shock. [8] The full climatic chamber dimensions are 75 ft x 40 ft x 24 ft. However, only half of the chamber was used for this testing. To simulate the weather conditions experienced by a New Jersey school bus during the course of a school year,[9] tests were conducted at four different ambient temperatures (20°F, 40°F, 65°F and 85°F) and relative humidity ranging from 37 to 90%. In total, 34 tests were performed.

Each test was approximately one hour in length. During this period, the temperature and humidity were held constant in the environmental chamber. After 1 hour, the time rate of change of the oil temperature was generally 0.2°F per minute. "Steady state" values of emission levels and fuel consumption were obtained by computing time averages over the final 10 min.

For the mobile study, a drive cycle was created using actual school bus routes. These routes were taken using GPS data from 5 different school districts. The school districts chosen represent rural, urban, and suburban conditions. Using this GPS data, a composite school bus cycle was created. The Rowan University Composite School Bus Cycle (RUCSBC) is repeated three times for each fuel type. To use the RUCSBC, engineers at Sensors, Inc. worked with Rowan University to develop software to enable a driver to follow a prescribed drive cycle on the PC screen. The target vehicle speed vs. time curve from the prescribed drive cycle is graphically displayed as a line along with the actual vehicle speed, which is shown as a large dot. A digital display is also included that shows the vehicle's actual speed, target speed, time elapsed and remaining, as well as total drive cycle time. Figure 1 below illustrates the RUCBSC.

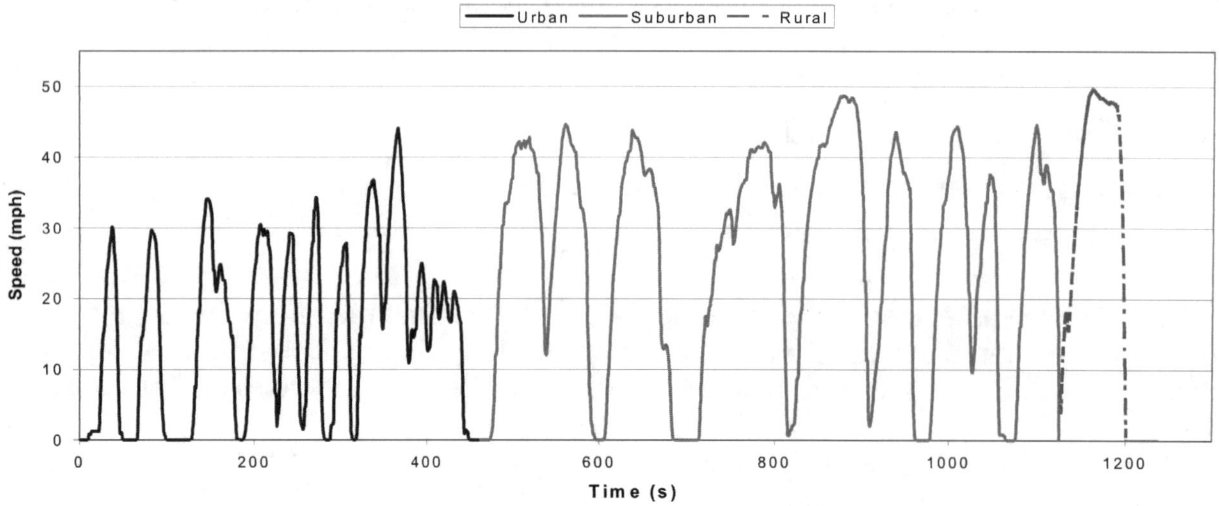

Figure 1: Rowan University Composite School Bus Cycle Separated by Representative Regions

Using the RUCSBC, four fuel types were studied. These fuel types are shown below in Table 2. Each fuel was tested three times in order to demonstrate repeatability.

Table 2: Test matrix for each school bus

Test	Fuel Type	Fuel Makeup
1	#2 Petroleum Diesel	Low Sulfur (~360 ppm) diesel)
2	B20	20% by Volume Biodiesel, 80% by Volume #2 Low Sulfur (~360 ppm) diesel)
3	ULSD	Ultra Low Sulfur Diesel (~15 ppm)
4	B20/ULSD	20% by Volume Biodiesel, 80% by Volume Ultra Low Sulfur Diesel (~15 ppm)

Results-Idle Emissions

From the results of the study, the average fuel consumption rate for idle is 0.5 gallons per. Because school buses do not have air conditioners and or electronic accessories, this value holds constant for the range of temperatures tested.

In addition to the fuel costs associated with idling, exhaust gas contains a variety of chemical species that have adverse effects on human health and the environment. The health risks associated with school bus idling are of a particular concern since much of the idling occurs with children on or near the bus. Particulate matter caused by incomplete combustion is known to contain carcinogens.[10,11] Also, NO_X CO and CO_2 are criteria air pollutants responsible for ozone depletion and global warming.

Experiments were conducted on each of the three buses over a temperature range of 20 to 85°F and relative humidity range of 37 to 90 F. Measurements of CO, CO2, NO_X, HC, PM and fuel consumption were made during idling conditions. In all, a total of 33 experiments were conducted.

CO_2 emissions are a direct function of fuel consumption. Generally, as more fuel is consumed, CO_2 emissions increase linearly. For the idle tests, the CO_2 emissions do not seem to demonstrate any type of pattern when temperature and humidity are changed. Similar results were found for the other two buses tested.

CO emissions are caused by incomplete combustion. With increasing ambient temperature, cylinder temperature also increases. This rise in cylinder temperature allows for a more complete combustion of the fuel. This results in less CO emissions and more CO_2 emissions. This is demonstrated in Figure 2.

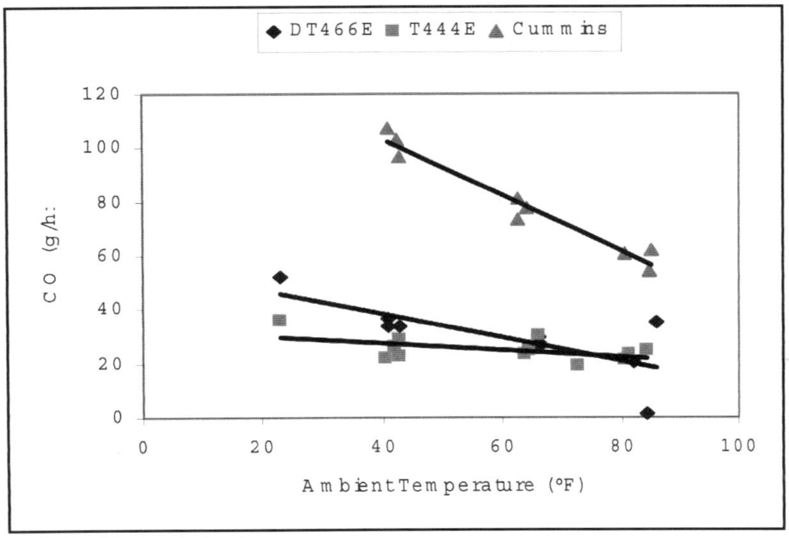

Figure 2: Effect of Ambient Temperature on CO Emissions.

Figure 3 shows the NO_X emissions in g/hr from DT466E engine. Emissions vary not only with temperature and humidity, but also with engine type. The Cummins B Series engine produced the highest NO_X emissions of the three buses tested. It should be noted that the Cummins engine is the oldest engine tested and does not contain an ECM. The NO_X emissions shown below demonstrate the effect of humidity on NO_X emissions. As expected, the results showed that for each of the buses tested, as humidity increases, NO_X emissions decrease. Since a higher inlet temperature should result in a higher cylinder temperature, it might be expected that NO_X would increase with increasing inlet temperature due to the well-known thermal NO_X mechanism. However, for each of the three buses tested in this study, NO_X emission generally was found to decrease with increasing ambient temperature.

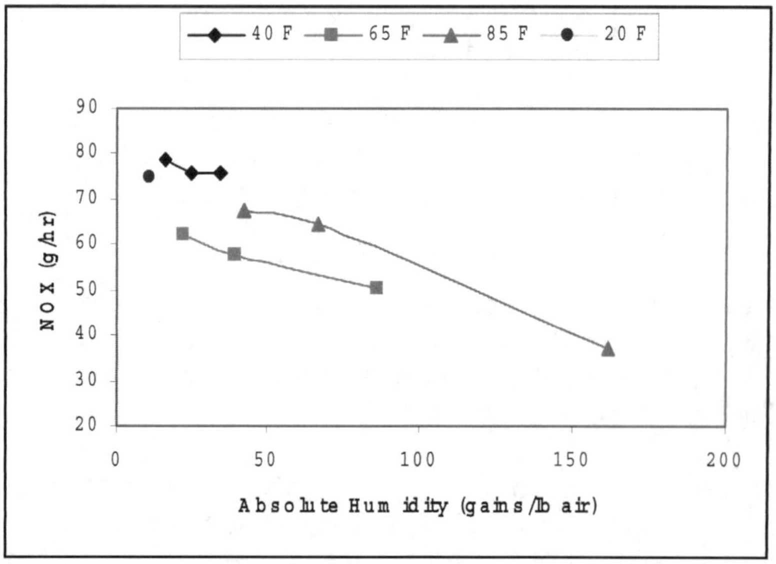

Figure 3: NO_x Emissions During Idling Conditions for School Bus with DT466E Engine.

Figure 3 also shows a decrease in NO_X emissions as humidity increases. For the DT466E, the measured NO_X emissions were slightly lower than those measured for the T444E engine. The DT466E engine showed a slight increase in NO_X when the ambient temperature was increased from 65 °F to 85 °F.

From the 34 experiments performed, there were no conclusions that could be made from ambient temperature or humidity on particulate concentrations. Surprisingly, the Cummins B Series Bus emitted 80% less particulates than the two Electronically Controlled school buses. The trade off is the higher CO_2, CO, and NO_X Emissions. The Cummins school bus produced 25% higher CO_2 emissions, 60% higher NO_X emissions and 70% higher CO and HC emissions. The results are shown below in Table 3.

Table 3: Average Particulate Matter for the Three Bus Engines Tested

Engine	PM (g/hr)
International DT466E	1.39
International T444E	1.11
Cummins B Series	0.27

Results-Fuel Type Mobile Testing

The fuel testing for the T444E is shown below in Figure 4. From the graph it is apparent that the use of alternative fuels lowers CO emissions. Since NO_X is a function of temperature and humidity, similar NO_X readings are measured for the 4 fuel types.

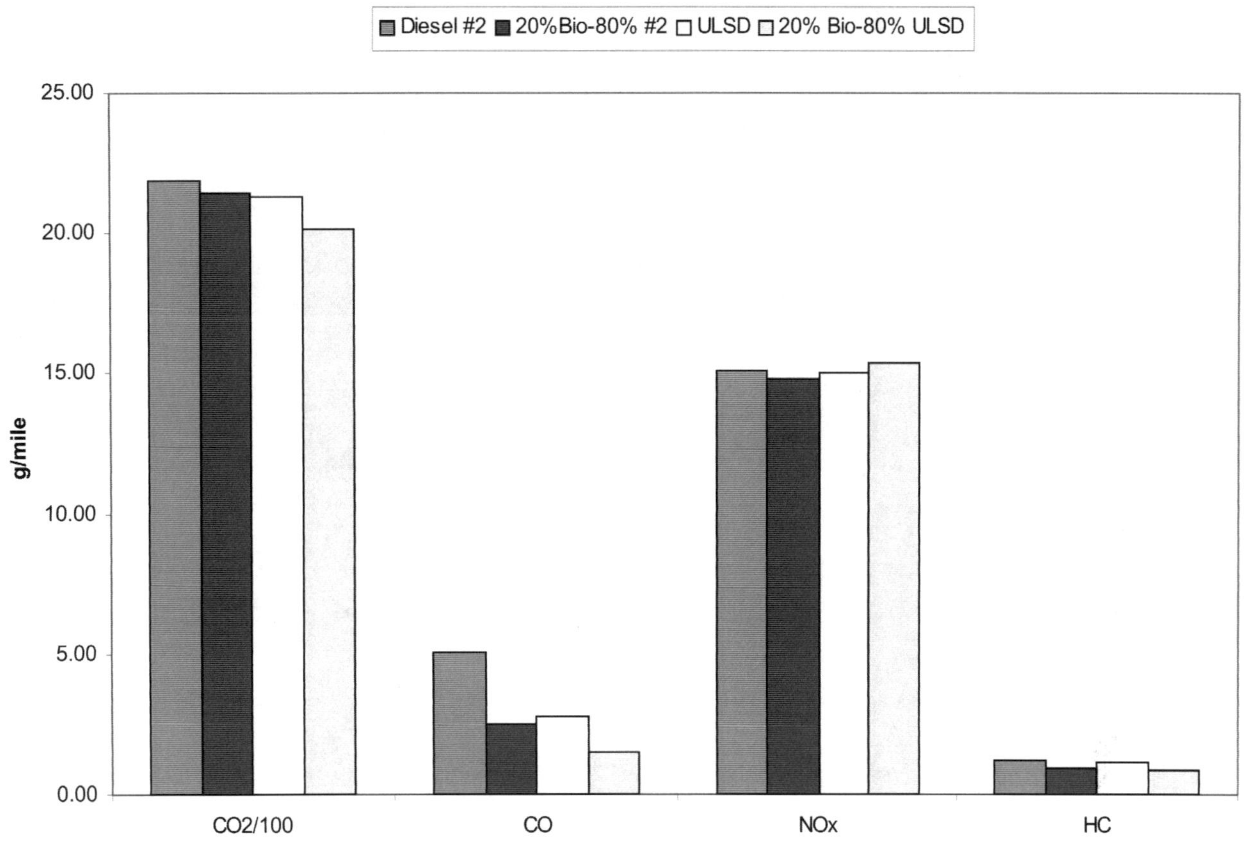

Figure 4: Mobile Fuel Testing Results Summary for the T444E

Conclusions

New Jersey is continuing its efforts to meet the National Ambient Air Quality Standards (NAAQS) by investigating emission reduction strategies for school buses. Little data has been collected on school buses in previous studies. In order to quantify the idling emissions, 33 tests were performed in an environmental chamber and 36 tests were performed on a test track using a cycle created from GPS data from actual school bus routes.

There are currently approximately 11,000 school buses in use in the state of NJ. Based on the idle emissions measurements presented herein (assuming an average idle time of 1 hr/day), the current NJ school bus fleet produces approximately 9.3 million kg CO_2, 90,000 kg CO, 150,000 kg NO_x, 35,000 kg HC and 2145 kg PM per school year.

In terms of the financial burden, a one-hour reduction of idling time per school bus per day for each of the 11,000 buses in NJ would result in an annual statewide cost savings of approximately $1.5 million. This calculation is based on an average school bus fuel consumption of 0.5 gallons of diesel fuel per hour and a price of $1.50 for #2 diesel[12].

In addition, the following conclusions were also found for the three school buses tested in this study:

- As ambient temperature increases, CO emissions decrease due to a more complete conversion of CO to CO_2.

- Particulate matter (PM) measurements did not show any relationship between temperature, humidity or fuel consumption.

- ULSD had little or no effect on HC, NO_x, or CO_2 emissions, however significantly decreased CO emissions

- The alternative fuels tested in this research alone will not provide the reductions needed to meet new emission regulations without engine modification. However, alternative fuels provide the ability to use other emission reduction devices that when used in conjunction with the fuels may meet new standards

ACKNOWLEDGMENTS

The authors wish to acknowledge the support of the New Jersey Department of Transportation. The authors also gratefully acknowledge the test engineering support of Todd Morris and Randy Babcock at the Aberdeen Test Center in Aberdeen, Md.

References

1. "Survey of School Buses in NJ", Polk Automotive Intelligence of Detroit, March 2002.
2. Lim, Han "Study of Exhaust Emissions from Heavy Duty Diesel Trucks and Commercially Available Idle Reducing Devices", SAE 2003-01-0288.
3. Pekula, N, B. Kuritz, J. Hearne, A. J. Marchese, R. P. Hesketh, "The Effect of Ambient Temperature, Humidity, and Engine Speed on Idling Emissions from Heavy-Duty Diesel Trucks," SAE 2003-01-0290.
4. Assessment and Standards Division Office of Transportation and Air Quality U.S. Environmental Protection Agency, "A Comprehensive Analysis of Biodiesel Impacts on Exhaust Emissions", EPA420-P-02-001, Oct. 2002.
5 Toback, A.T., Hearne, J.S., Hesketh, R.P., Marchese, A.J., The Effect of Ambient Temperature and Humidity on Idling Emissions from Diesel School Buses", SAE World Congress, #2004-01-1087
6. "SEMTECH-D Operator's Manual." Revision 2.1, Sensors, Inc.
7. "PM-300 Operator's Manual." Revision 5.1, Sensors, Inc.
8. Aberdeen Test Center Facilities/capabilities Guide, http://www.atc.army.mil/fac_guide/ facilities/envcham4.html, August 21, 2002.
9. National Oceanic and Atmospheric Administration- National Environmental Satellite and Information Services http://nndc.noaa.gov/?http://www.nndc.noaa.gov/cgi-bin/nndc/buyOL-002.cgi?FNC=sendstation__Ausa.
10. IARC(International Agency for Research on Cancer) Diesel and Gasoline Engine Exhausts and some Nitroarenes. IARC, Lyon, Frankce, ISBN 92 832 1246 0, 1998.
11. Groves, J. and J.R. Cain, "A Survey of Exposure to Diesel Engine Exhaust Emissions in the Workplace," Ann. Occup. Hyg., 44(6) pp. 435-447 (2000).
12. US Department of Energy. "Gasoline and Diesel Fuel Update. http://tonto.eia.doe.gov/oog/info/gdu/gasdiesel.asp, accessed August 16, 2003.

Paper T4003-25a

Developing Sustainable Process Technology

Joseph A. Kocal
UOP LLC
Des Plaines, IL 60017-5017

Prepared for Presentation at AIChE Spring Meeting, April 25-29, 2004

Abstract

The key to sustainable growth is innovative technology that addresses performance elements for economic growth, environmental protection, and social progress. Environmental regulations, the push for social progress, and profitability demands will drive refining, chemical, and other industries to meet this innovation challenge. Near-term efforts focus on process optimization, energy utilization, and emissions reduction using incremental technology improvements. In the future, hazard reduction, feedstock substitution, and new energy sources will become more critical. This paper describes approaches to developing sustainable technology and provides pertinent examples in areas such as direct natural gas utilization, desulfurization, and elimination of hazardous liquid acids as catalysts.

Introduction

There are many definitions for sustainable development in use today. The World Resources Institute definition: "Growth that meets the economic, social, and environmental needs without compromising any of them" encompasses most concepts. The chemical, petrochemical, and petroleum refining process industries have slowly evolved toward maintaining sustainable development. Although the progress towards sustainable development has been significant, it has largely been directed at increasing efficiency, minimizing waste, and eliminating harmful emissions for existing processes. Much of this has been driven by the savings resulting from more efficient operations and government legislation. For example, Dow Chemical has utilized new recovery systems and more efficient processing equipment to reduce emissions by 43% and waste by 30%. This resulted in a 180% return on investment at the Midland, Michigan facility.[1] Improvements in technology such as this will extend our resources, but in the long run are not sustainable.

Although there will continue to be advances in more efficient and cleaner operation of existing processes, there is clearly a need to be proactive and develop technology based on the tenet of strategic development. A long-term strategic plan is necessary to successfully move forward to the development of sustainable technologies. The goal of creating the most efficient processes and better products, with minimal use of resources and generation of waste and pollution, is common sense and good business. Identifying what is truly sustainable will be a moving target as complexities in life cycle analysis are better understood. As a result, the strategies to achieve sustainable development will be modified. UOP does not have the answers to these complex issues. At this time we are developing a vision for sustainable development and plan to address the critical issues in a logical sequence.

The challenges to creating sustainable technology are numerous and the magnitude daunting. Hart has estimated it will be necessary to grow the economy tenfold to sustain a doubling of the population.[2] In order to maintain the current environmental standard, he estimates it will require approximately a twenty-fold improvement in technology.

Innovation is the enabler to succeed in developing technology that not only maintains, but exceeds, current quality of life standards and dramatically improves our environment. Throughout the history of the process industries, innovation has been the key to success. Breakthrough innovations have enabled step changes in performance in the past. The response to fuel quality challenges such as the elimination of lead and the development of ultra-low sulfur fuels are examples. The identification of a solid catalyst in the Detal™ process for production of linear alkylbenzene used in preparation of a biodegradable surfactant allowed for the elimination of hazardous hydrofluoric acid in a cost-effective manner. No new HF-based processes for this technology have been built since the introduction of Detal. These discoveries improved the environmental situation and local safety in an economical way. However, these new technologies are clearly not sustainable as they consume irreplaceable resources. Two examples of sustainable technology based on renewable resources are the commercialization of polylactic acid (PLA) from corn starch by Cargill Dow LLC and 1,3-propanediol (PDO) from glucose by Dupont.[3, 4] Although these 2 processes are based on renewable resources, it is not clear that they will be sustainable. Only time will tell. UOP believes breakthroughs such as these as well process innovations utilizing renewable feedstocks will make sustainability objectives achievable.

Developing Sustainable Technology

UOP is in the process of developing a vision for sustainable growth. We have a developed a preliminary road map (Figure 1) to reach this goal. This map is an initial proposal and indicates areas we feel may be important in developing sustainable technology. UOP will be clarifying its vision, developing an improved road map and plan to achieve sustainable development as time progresses and will work with its customers to ensure that social, environmental, and economic needs of all involved are satisfied.

There are progressive stages in the development of sustainable growth that span from optimization efforts to the ultimate goal of a society that utilizes renewable resources. The Y-axis on the figure indicates the degree of technical difficulty that is increasing with time. The exact contribution to sustainable development made by each of the topics outlined in the figure needs to be determined. Additional areas will be identified and others deemphasized or dropped as we learn more. We are beginning to put together information related to these topics to determine their relevance to sustainable growth. As expected, there is currently substantially more effort in the early stages along this journey, but increasing realization that we must focus our research and development efforts on the latter stages. A detailed description of the four stages to sustainable development with specific examples follows.

Figure 1. Sustainable Technology Road Map

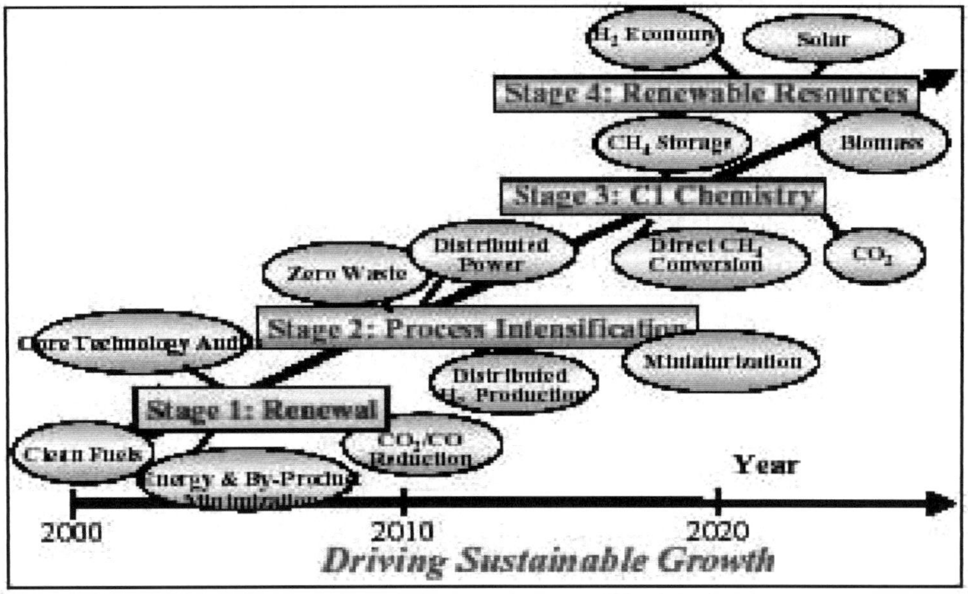

Stage 1 of this process is renewal. Past and continuing efforts have primarily been focused in this stage of development. Activities include improving feedstock utilization, driving down energy intensity of existing processes, developing clean fuels, minimizing undesirable by-products, and reducing environmental impact of existing processes while maintaining profitability. This is the first step toward sustainable development.

UOP and others have numerous efforts in this stage. Some of the programs include desulfurization technology from hydroprocessing to polishing off-spec product to meet "zero" sulfur specifications. Other programs involve upgrade of LCO feedstock to be utilized as a clean fuel or as a petrochemical feedstock. Additionally, UOP is continually looking at ways of managing its catalyst products for the various processes. This includes improved catalyst production yields and catalyst formulations that are environmentally compatible from cradle to grave.

An example of UOP technology in this stage is the Alkylene™ process for the production of motor fuel alkylate gasoline. Currently, alkylate gasoline is produced via the reaction of isobutane with C_3-C_5 olefins in the presence of either hydrofluoric or sulfuric liquid acids. The alkylate gasoline prepared by this route is almost completely paraffinic and is an excellent blending stock because it has essentially no sulfur, nitrogen, aromatics, olefins, or volatile organic compounds. In addition, it has acceptable vapor pressure, boiling range, and high octane number. The combination of these properties makes alkylate gasoline a preferred component for our transportation fuel. It would be desirable to increase the amount of alkylate in our gasoline pool and minimize or eliminate the undesirable components. Development of a solid acid catalyst to replace these hazardous liquid acids is highly desirable. This process would be inherently safer and would not pose the environmental risk that currently exists if accidental release of acid occurs. Although this risk is generally localized, there is significant transport of HF to make up that lost to process by-product. This heavy fluoride-containing by-product is generally burned resulting in unwanted emissions. The total amount of hydrofluoric acid on site at refineries worldwide is approximately 3.3 MM Kg or 0.87 MM gallons. Additionally, the operation using hydrofluoric acid results in the creation of a large amount of solid waste as a variety of process streams must be treated to remove organic and inorganic fluorides (Table 1).

Table 1. World-Wide Cost of Clean-Up and Make-Up in HF Alkylation.

Material	Amount, MM lbs/yr	Cost, $MM/yr
Alumina	9.8	5.9
KOH	3.9	2.0
Lime	5.9	1.9
HF Acid Makeup	33.2	23.4
TOTAL	52.8	33.2

Note: Estimate made on 975,000 BBL/day alkylate production world-wide

Over time the operating efficiency of hydrofluoric acid alkylation units has improved in order to minimize waste streams as well as minimize the actual hydrofluoric acid inventory. However, the only way to eliminate these issues completely is to develop a process based on a heterogeneous acid catalyst. Such a process would eliminate the hydrofluoric acid transportation and inventory, result in no waste polymer product, and would not require fluoride scrubbing and concomitant disposal of fluoride-containing solids.

Technologically, the Alkylene process is a break-through. Several significant inventions were required to make it technically feasible. The development of a unique solid-acid catalyst and transport reactor by UOP allows for the potential elimination of hazardous liquid acid processes. About 1 MM gallons of hydrofluoric acid inventory could be eliminated, transport of 33 MM lbs (4 MM gallons) of hydrofluoric acid per year would be stopped, and ca. 20 MM lbs per year of other fluoride containing solids would not have to be land filled.

The costs for the new and sustainable technology are found to be similar to those of the existing technology (for an 8,000 BBL/day plant, an EEC of $MM 43.5 for the Alkylene process and $MM 40.5 for hydrofluoric acid alkylation). The increased costs for the novel Alkylene reactor are offset by costs for HF mitigation capital and metallurgy in parts of the unit. With maturity and experience, the costs for this new technology are expected to decrease. UOP is currently developing an improved catalyst and process that will further reduce the cost of the Alkylene process thereby making it more economically competitive with hydrofluoric acid technology.

Stage 2 of UOP's roadmap entails process intensification which involves substantial simplification to minimize process steps, reduce capital costs, and improve product yields. Early efforts in process intensification included combining process steps such as reaction and separation. Reactive distillation has been effectively demonstrated in commercial environments for methylacetate and MTBE production, respectively.[5,6] In these cases separation during reaction also allows high conversion to be achieved by eliminating the equilibrium constraints at process conditions. This process intensification via combining process steps usually leads to lower capital and operating costs and many time improved yields, more efficient use of feedstock, minimization of by-products, and lower energy requirements. Again it is a step in the right direction, but not sustainable.

Process intensification has recently involved miniaturization of the reactor in order to greatly reduce heat and mass transfer resistances. Dramatic changes in technology can occur with the advent of order of magnitude increases in the rates of heat and mass transfer. Distributed production of power and chemicals would be realized as a direct result of effective miniaturization. Process intensification would allow precise control of reaction conditions that could dramatically impact product yields and purity. These improvements lead to reduced use of raw materials, decreased cost of purification and waste handling, and lower energy consumption. Because total volume is dramatically reduced, difficult to produce or hazardous chemicals can be produced as needed and in a distributed way at point of use. This minimizes the inherent handling problems or environmental and safety issues resulting from transportation of these materials. Distributed production of hydrogen is a prime example of this miniaturization for distributed production. In the Hythane^{TM} process, mini-reformers are used to convert methane to hydrogen or a hydrogen/natural gas blend for fuel use in cities around the world, including Chicago, Los Angeles, and Palm Springs.[7]

Miniaturization might also be used in the distributed production of chemicals. Two examples include the safe production of hydrogen peroxide from hydrogen and oxygen or the production of chemicals or fuels from synthesis gas. More detailed study into the potential improvements in product yields, waste minimization, and costs are required to fully understand the impact of miniaturization. At this time it offers the potential to produce hazardous materials or conduct potentially dangerous oxidation reactions on a small scale that is inherently safer.

Stage 3 shifts the source of hydrocarbon for transportation and chemicals to methane. Methane currently is reformed at elevated temperatures and pressures to synthesis gas. This mixture of hydrogen and carbon monoxide can then be converted via the well-known technologies of methanol synthesis and Fischer-Tropsch synthesis to eventually produce a variety of chemicals and fuels. In this stage, focus will be given to direct methane conversion technologies that will substantially reduce dependence on petroleum and coal. It will also provide substantial reductions in energy and emissions resulting from the high temperature indirect routes to power, chemicals, and fuels involving synthesis gas. Storage of methane for transport or use on demand is also a prime target for development in the future. A much more detailed analysis of the potential for direct conversion approaches to replace indirect and established routes for production and utilization of synthesis gas must be completed. A hypothetical process is considered below for discussion. At this time UOP is considering its opportunities in the area of direct conversion of methane to chemicals or fuels.

The direct catalytic conversion of methane has been actively pursued for many years. Much of the emphasis has been on the direct production of methanol via selective partial oxidation[8], coupling of methane to ethylene[9], or methane aromatization[10]. At this time none of these technologies has been demonstrated commercially due to low yields of desired products due to combustion by-products or low equilibrium conversion at reasonable process temperatures and pressures. The potential benefits of a hypothetical process for the direct partial oxidation of methane to methanol[11] are presented as an example.

According to data from 2001, methanol is consumed at a rate of about 67 billion pounds per year and is the eighth ranked chemical in production value in the United States. Current energy usage for production is 0.015 MMBTU/lb with nearly 950 trillion BTU energy per year consumed globally for methanol production. About 80 percent of the methanol is produced from natural gas. Therefore, a more efficient and direct process should result in dramatic energy savings and eliminate substantial amounts of CO_2 emissions. Table 2 lists the benefits of a hypothetical process for direct methanol production. Assuming a modest methanol yield of 9.5% (10% conversion and 95% selectivity), it was calculated that raw material costs decrease from 0.74 to 0.61 cents/lb, operating costs reduce by 50% (0.52 to 0.26 cents/lb.), and total cost is decreased from 3.06 to 1.83 cents/lb. Additionally, it is estimated that capital investment could be reduced by as much as 50%. Most importantly, this new process could result in a net 17 trillion BTU/yr savings. Development of a process with further improved yields maintaining high selectivity will produce greater energy and economic savings. The validity of these estimates will need to be checked as process development progresses.

Table 2: A Cost Comparison between of Methanol Produced by Conventional Technology in the Mid/Far-East and a Proposed Process

	Conventional Process (cents/lb)	**Hypothetical Methane to Methanol Process (cents/lb)**
Raw Material Cost	0.74	0.61
Catalyst and Chemicals	0.06	0.08
Operating Cost	0.52	0.26
Tax/overhead	0.50	0.25
Depreciation	1.25	0.63
Cash Cost	1.82	1.20
Total Cost	3.07	1.83

Table 3 lists the environmental benefits. Substantial reductions in oxide and VOC emissions will be realized upon commercializing such a process. Furthermore, development of a process with improved yields maintaining high selectivity will produce greater energy and economic savings and further minimize pollutants. It is important to note this hypothetical process is still a long way from commercialization and several technological breakthroughs in catalyst and process design will be required to realize these benefits.

Table 3. Methane to Methanol: Calculated Reduction in Pollutants (Assumining Technology is Commercialized in 2012)

POLLUTANT REDUCTIONS (lbs)	2005	2010	2015	2020
Carbon (MMTCE/yr)	N/A	N/A	0.026	0.082
Nitrogen Oxides (NOX)	N/A	N/A	405,463	1,269,365
Sulfur Oxides (SOX)	N/A	N/A	360,238	1,127,782
Carbon Monoxide (CO)	N/A	N/A	79,533	248,991
Volatile Organic Compounds (VOCS)	N/A	N/A	8,811	27,584
Particulates	N/A	N/A	6,628	20,749
Other (million lbs)	N/A	N/A	130	407

Stage 4 moves to the utilization of renewable resources. As technology moves well into this stage, the world's energy will come from primarily renewable resources, such as solar, biomass, and thermo chemical. The desired fuel would be hydrogen which is ultimately derived from water and not a hydrocarbon. Efficient hydrogen storage will also have to be developed for use by this time.

Considerable research effort is being placed into the development of renewable resources. Although steady progress is being made in areas such as photovoltaics, there are numerous discoveries yet to be made. At this time, the cost of producing electricity via electrolysis of water using photovoltaics is about $0.12/kWh. In order to be competitive, this cost must be reduced substantially via improved efficiency of the overall system and reduced cost resulting from mass production.

The utilization of biomass feedstock is in most cases not economically competitive today. One process that is commercially viable is the production of polylactic acid (PLA) from corn starch.[3] Although lactic acid production from the fermentation of carbohydrates has been known for over 150 years, processing the lactic acid into useful polymers was a difficult and expensive operation. Cargill's primary technical innovation was the development of melt-phase technology for the ring opening polymerization of the lactide dimer. Upon demonstration of the technology at the pilot scale, Cargill teamed with Dow Chemical to provide leverage into the polymer market. Cargill Dow LLC was formed in 1997 to commercialize the PLA technology. Its PLA polymer products are 100% derived from renewable resources and are produced with the cost and performance necessary to compete with petroleum-based packaging materials and traditional fibers. Detailed life-cycle assessments of PLA compared with petrochemical derived polymers have been performed and published; net fossil fuel reduction of 20-50%, with 15-60% reduction in the generation of greenhouse gases is claimed.

For economic viability of biomass-derived processes to be achieved, innovations that produce high added-value chemicals will likely be required. For use as a source of fuels or bulk chemicals, the cost of collecting and processing the biomass, as well as improved conversion technology, will need to be developed. As the availability of oil and gas eventually decreases and the cost increases, technology involving use of biomass should become more important without the need of government subsidies. At this time UOP is considering where it can apply its core compencies in the area of renewable resources. During this study period we are trying to identify which potential technologies will truly lead to sustainable development. This is a difficult analysis requiring detailed life cycle studies.

Conclusion

Sustainable growth is a major issue facing the chemical, petrochemical, and refining industries in the foreseeable future. The challenge to meet environmental, societal, and economic performance simultaneously will require both breakthrough and incremental innovations. The industry has met such challenges successfully in the past. At UOP, we are convinced that by delivering innovation in process technology and equipment, catalysts and advanced materials, as well as solutions and services we will contribute to developing a sustainable future. This effort will be conducted with the input of our customers to ensure we are meeting their needs.

References

1. Parker, M.D., November 5, 2001, "Pressing on with Our Journey-The Imperative of Sustainable Development", CAEC.
2. Hart, S.L., January-February 1997, "Strategies for a Sustainable World", Harvard Business Review.
3. Drumright, Gruber, Henton, "Polylactic Acid Technology", Adv. Mat. 2000 12(23), 1841-1846.
4. a) US Patent 5,686,276, 1995 "Bioconversion of a fermentable carbon source to 1,3-propanediol by a single microorganism" b) US Patent 6,514,733, 2003 "Process for the biological production of 1,3-propanediol with high titer".
5. Siirola, J.J., An Industrial Perspective on Process Synthesis. AIChE Symp. Ser. 1995, 92 (304), 222.
6. Davis, S., "UOP Ethermax Process for MTBE, ETBE, and TAME Production", Handbook of Petroleum Refining Processes - second edition, 1997, 13.9-13.12.
7. US Patent 5,139,002, 1992 "Special purpose blends of hydrogen and natural gas".
8. Zhang, Q, He, D. Zhu, Q., Recent Progress in Direct Partial Oxidation of Methane to Methanol, Journal of Natural Gas Chemistry, 12 (2003), 81-89.
9. Lunsford, J.H., The Catalytic Oxidative Coupling of Methane, Angew. Chem. Ed. Engl., 34 (1995), 970.
10. Echevsky, G.V., Kodenev, E.G., Kikhtyanin, O.V., Parmon, V.N., Direct Insertion of Methane intoC3-C4 Paraffins Over Zeolite Catalysts: a Start to the Development of New One-Step Catalytic Processes for Gas to Liquid Transformation, Applied Catalysis A: General, 258 (2004), 159-171.
11. Maher, J.M., Development of Novel Catalysts for the Oxidation of Methane to Methanol with High Selectivity and Throughput, Proposal for the Chemicals and Forest Products Industries of the Future, Solicitation DE-PS36-03GO93015, January 2004.

Novel Membranes and Membranes Processes for Industrial Separations - Part II

J. Wijmans, K. Amo, R. Baker, R. Blanc, A. Da Costa, R. Daniels, J. He, A. Ariwala and I. Pinnau*
MTR, Inc.
1360 Willow Road
Menlo Park, CA USA, 94025
Tel. (650) 328-2228 x 118
Email: wijmans@mtrinc.com

Abstract

Separations are an indispensable part of petrochemical and refinery operations. Traditional separation processes, such as distillation, pressure swing adsorption and cryogenic condensation, are still dominant. However, the development of novel membranes and membranes processes is increasing the potential of membrane separation processes in these industries.

Over the years past, MTR has executed a number of research and development projects focused on membrane applications in refineries and in petrochemical plants. Most of these projects received financial support from the Office of Industrial Technologies (Program Office of Energy Efficiency and Renewable Energy, U.S. Department of Energy) through its Chemical Industry of the Future initiative and the Small Business Innovation Research Program.

In this presentation we will give an overview of the research and development activities listed above. The applications to be discussed are:

1. Olefin/paraffin separations
2. Hydrogen/hydrocarbon separations
3. Isomer separations

NOVEL MEMBRANE REACTOR FOR FISCHER-TROPSCH SYNTHESIS

Michael C.J. Bradford, Mure Te and Allen Pollack*
CeraMem Corporation
12 Clematis Avenue
Waltham, MA 02453
Tel: (781) 899-4495
Fax: (781) 899-6478
** Corresponding Author*

ABSTRACT

A prototype catalytic membrane reactor for Fischer-Tropsch synthesis has been developed. Experimental data demonstrate that the product yield exhibited by a Pt-Co/γ-Al_2O_3 catalyst in the catalytic membrane reactor exceeds that observed by the same catalyst in a tubular, fixed-bed reactor.

1. INTRODUCTION

Fischer-Tropsch synthesis (FTS), i.e., the conversion of synthesis gas (H_2 and CO) to C_2+ hydrocarbons, has been used since the early 20th century to produce liquid fuels and chemicals from natural gas, coal, and petroleum coke. Although several reactor designs for FTS have been developed and commercialized,[1] other, potentially superior designs have been envisioned.[2] During the past few decades, the U.S. Department of Energy ("DOE") has sponsored numerous research and development programs in this area with the ultimate objective of employing the technology domestically in order to increase U.S. energy independence.[3] However, FTS technology is currently only economically viable at large scales,[4] such that more than 60% of the world's natural gas fields cannot likely be economically exploited using currently available FTS technology.[5] Therefore, there is a need to develop new FTS reactor and process technology that permit the economic utilization of small natural gas fields, including landfill gas fields.[6]

The U.S. DOE Small Business Innovation Research (SBIR) program provided CeraMem Corporation with Phase I and Phase II awards in 2000 and 2001, respectively, to utilize CeraMem's proprietary monolith and membrane technology [7] in order to develop a novel membrane reactor concept for FTS that could enable the economic monetization of small natural gas fields, including land fill gas fields. While the use of membranes in FTS reactors for the continuous extraction of H_2O has already been studied [8] and patented,[9] there are in principle numerous alternative three-phase membrane reactor configurations for FTS (such as that proposed by J.Akyurtlu et al.),[10] one of which CeraMem is developing in its current DOE SBIR program.

CeraMem's objective is to develop a unique membrane reactor design as part of an overall process for the conversion of natural gas or landfill gas to liquids (Figure 1) that incorporates some advantages of fixed-bed and slurry-bubble column technology, including:

- An attrition-resistant catalyst that is immobilized within the reactor, thus eliminating the need for unit operations dedicated to catalyst/wax separation;
- Efficient radial and axial gas mixing, resulting in relatively uniform gas concentrations and reaction rates throughout the entire reactor volume;
- Low axial pressure drop;
- Negligible mass transfer limitations;
- An efficient mechanism for heat management; and
- A potentially smaller footprint than competing reactor technologies which make it an attractive option for the monetization of off-shore natural gas [11] and landfill gas.

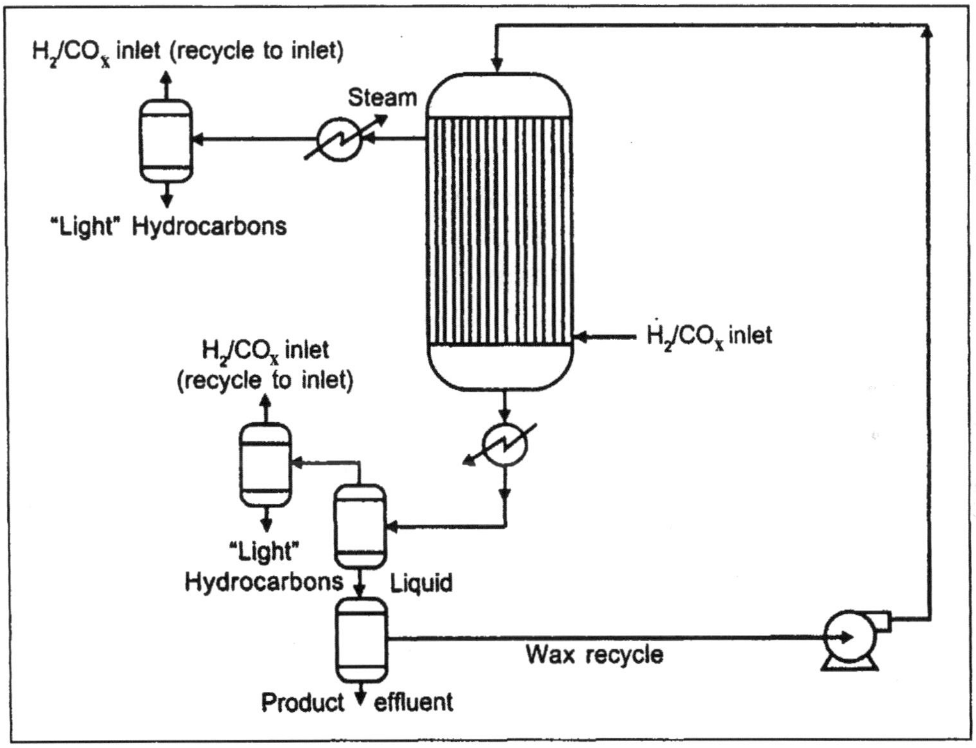

Figure 1. A simplified schematic of CeraMem's membrane reactor for Fischer Tropsch Synthesis.

2. EXPERIMENTAL

2.1. Catalyst Preparation and Characterization

Although several catalyst formulations identified in the journal and patent literature [12,13,14,15] were prepared, characterized and evaluated during this program, CeraMem eventually decided to focus exclusively on the use of Pt-Co/γ-Al_2O_3,[16] which was prepared using two sequential steps:

(1) 20 grams of $Co(NO_3)_2 \cdot 6H_2O$ (Aldrich, ACS Reagent Grade, 98+%) was dissolved in 25 ml of distilled water, and then 25 grams of calcined γ-Al_2O_3 (Sasol North America Inc., Catalox SCCa 5/150) was suspended in this solution. The slurry was dried over a period of ~2.5 h at 75°C in a rotary evaporator. Then, the dried material was removed from the rotary evaporator and calcined at 230°C for 2 h in flowing dry air (120 $cm^3/h \cdot g_{cat}$).

(2) A solution was prepared by adding 17.5 grams of $Co(NO_3)_2 \cdot 6H_2O$ and 0.42 grams of a tetraamine platinum(II) nitrate solution (3%Pt, Pressure Chemical Company) to 25 ml of deionized water. The calcined material from the prior synthesis step was added to the solution, and the slurry was vacuum treated at 75°C in a rotary evaporator until freely flowing. The recovered material was then calcined at 230°C for 2 hours in flowing dry air (120 $cm^3/h \cdot g_{cat}$).

Catalyst samples were characterized via:

- Direct current plasma emission spectroscopy (DCPES), to quantify catalyst composition;
- Nitrogen adsorption at 77 K, to determine the surface area and pore size distribution;
- X-ray diffraction (XRD) to determine the bulk crystalline structure;
- Hydrogen and oxygen chemisorption, to quantify the accessible cobalt surface area after reduction;
- Temperature programmed reduction (TPR), to quantify the extent of cobalt reduction and dispersion; and
- Scanning electron microscopy (SEM), to determine catalyst layer morphology after application to the tubular membrane elements.

2.2. Apparatus for Catalyst Evaluation

An apparatus was constructed for catalyst evaluation in either a fixed-bed or membrane reactor. Four mass flow controllers (Porter Instrument Company) are used to independently feed carbon monoxide (CO, 99.5%), hydrogen (H_2, 99.999%), carbon dioxide (CO_2, 99.99%), and nitrogen (N_2, 99.999%) to the feed-gas manifold. The inlet gas mixture is delivered to either the fixed-bed reactor (FBR) or the membrane reactor (MR) by using a three-way switching valve.

The FBR is constructed of type 316 stainless steel tubing (3/8" O.D) and is equipped with: an internal J-type thermocouple to monitor the catalyst bed temperature; a thermocouple on the outer surface of the reactor for temperature control; a pressure transducer; and a pressure relief valve. The MR is a custom-made reactor designed to house commercially available tubular membranes (such as those offered by Pall Corporation), and is equipped with: an internal J-type thermocouple to monitor the tube-side temperature; a pressure transducer to monitor shell-side pressure; a differential pressure gauge to measure transmembrane partial pressure; and a pressure relief valve. The temperature of each reactor is maintained through the use of electrical heating tape using programmable temperature controllers, and the system pressure is maintained through use of a backpressure regulator (TESCOM).

The effluent from either reactor is directed through a heated transfer line (typically maintained at 180°C) to a "hot trap" maintained at 110°C and then to a "cold trap" maintained at *ca.* 0.1°C through use of an external cooling jacket in which a water-ethylene glycol mixture is circulated. The effluent from the cold trap is directed to either a vent or to the injection valve of an on-line gas chromatograph for analysis (see Section 2.5).

2.3. Procedures for Catalyst Testing in the Tubular Membrane Reactor

Calcined catalyst was dispersed in a proprietary mixture, ball-milled, and then either dried for subsequent use in the fixed-bed reactor (Section 2.4) or coated onto the surface of a tubular membrane element. In this program, commercially available membrane tubes (Pall Corporation) were cleaned and modified prior to application of the catalyst layer. After application of the catalyst layer, the catalytic membrane tube was calcined in flowing dry air at 450°C, cooled to room temperature, installed in the membrane reactor housing, reduced *in-situ* with flowing H_2 (2,330 $cm^3/h \cdot g_{cat}$) for 5 h at 450°C, cooled to the desired reaction temperature, and then exposed to a $H_2/CO/CO_2/N_2$ mixture in order to initiate Fischer Tropsch Synthesis. Thereafter, the system pressure was increased to the desired operation condition.

2.4. Procedures for Catalyst Testing in the Fixed-Bed Reactor

Catalyst recovered from the dried "mixture" after ball-milling (see Section 2.3) was calcined in flowing dry air at 450°C, cooled to room temperature, ground and sieved (typically to a size fraction of 45-75 μm), dispersed in SiC (80 grit, Electro Abrasives Corporation) that had been calcined in flowing dry air at 450°C, loaded into a 3/8" OD stainless steel reactor tube between slugs of high purity quartz wool, reduced *in-situ* with flowing H_2 (2,330 $cm^3/h \cdot g_{cat}$) for 5 h at 450°C, cooled to the desired reaction temperature, and then exposed to a $H_2/CO/CO_2/N_2$ mixture in order to initiate Fischer Tropsch Synthesis. Thereafter, the system pressure was increased to the desired operation condition.

2.5. Procedures for Product Analysis

During catalyst testing in either the FBR or MR, gaseous effluent from the "cold trap" was directed to an on-line SRI gas chromatograph (GC) for analysis. A 15 ft 60/80 Carboxen 1000 packed column and thermal conductivity detector were used to quantify CO, CO_2, H_2, N_2, and CH_4, and a 30m MXT-QPLOT capillary column and flame ionization detector were used for C_1-C_8 gas-phase hydrocarbon analysis.

The "cold trap" and "hot trap" were drained periodically during, and upon the completion of, each experiment in order to collect product and check the overall material balance. The aqueous phases were analyzed via GC to identify and quantify dissolved oxygenates (Atlantic Analytical Laboratory, Inc.), and the hydrocarbon (liquid and solid) phases were analyzed via GC to identify and quantify the hydrocarbon distribution (Baseline DGSI). Gas-phase and liquid-phase product analysis were thereafter combined in order to generate complete Anderson-Shulz-Flory plots.

3. RESULTS

3.1. Catalyst Characterization

The elemental composition of the calcined Pt-Co/γ-Al_2O_3 catalyst as determined via DCPES (0.05wt%Pt and 20.5wt%Co) is in good agreement with the nominal composition (0.04wt%Pt and 21.5wt%Co). Surface areas, pore volumes and average pore sizes of calcined γ-Al_2O_3 and Pt-Co/γ-Al_2O_3 are provided in Table 1. Through a combination of TPR and chemisorption experiments it was determined that after the calcined Pt-Co/γ-Al_2O_3 catalyst is exposed to H_2 for 10 hrs at 350°C, 58% of the Co_3O_4 is reduced to Co^0, and the Co^0 dispersion is 8.2%, such that the average Co^0 cluster diameter for the Pt-Co/γ-Al_2O_3 catalyst is approximately 12.6 nm (Table 2). Line-broadening analysis of the calcined Pt-Co/γ-Al_2O_3 catalyst indicated that the volume-weighted cluster size of Co_3O_4 is approximately 13.4 nm. The expected volume-weighted crystallite size of Co^0 after reduction of these clusters is approximately 10.6 nm (based on the differences in density and crystal structures of Co_3O_4 and Co), in reasonable agreement with the surface-weighted Co^0 particle size obtained via chemisorption (12.6 nm).

Table 1.

Surface areas, pore volumes and average pore sizes of γ-Al_2O_3 and Pt-Co/γ-Al_2O_3.[a]

Material [Particle Size Fraction]	History	Surface Area (m^2/gram) [b]	Pore Volume (cm^3/gram) [c]	Mean Pore Size (nm) [c]
γ-Al_2O_3	Calcined	126 ± 8	0.465 ± 0.007	9.7 ± 0.2
Pt-Co/γ-Al_2O_3 [<45 μm]	Calcined	86.4 ± 4.4	0.230 ± 0.014	7.98 ± 0.04
Pt-Co/γ-Al_2O_3 [<45 μm]	Ball-Milled & Calcined	91.2	0.20	7.92

[a] Non-biased standard deviations ($\pm 1\sigma_{n-1}$) are based on independent measurements of different samples.

[b] Surface areas were calculated from 7-point isotherms obtained from $0.05 \leq P/Po \leq 0.50$ using the BET method.

[c] Pore volume and pore-size distributions were calculated from 40-point isotherms (including 20 adsorption and 20 desorption points) from $0.05 \leq P/Po \leq 0.995$ using the BJH method.

Table 2. Degree of reduction, dispersion and average particle size of Co^0 in the Pt-Co/γ-Al_2O_3 catalyst as estimated by H_2 desorption and O_2-reoxidation (assuming 23.3 wt%Co after reduction).

U_{irr} (H_2) (μmol H_2/g_{cat})	$D_{100\%}$ [a]	$d_{s,100\%}$ [a] (nm)	U_{irr} (O_2) (μmol O_2/g_{cat})	% Reduction	$D_{58\%}$ [b]	$d_{s,58\%}$ [b] (nm)
92.7	4.8%	21.7	1,513	58%	8.2%	12.6

[a] Average Co dispersion ($D_{100\%}$) and particle diameter ($d_{s,100\%}$) calculated assuming 100% reduction.
[b] Average Co dispersion ($D_{58\%}$) and particle diameter ($d_{s,58\%}$) calculated assuming 58% reduction.

In addition, scanning electron microscopy (SEM) was utilized to examine the surface morphology and cross-sectional structure of catalyst-coated, tubular membrane elements (Figure 2).

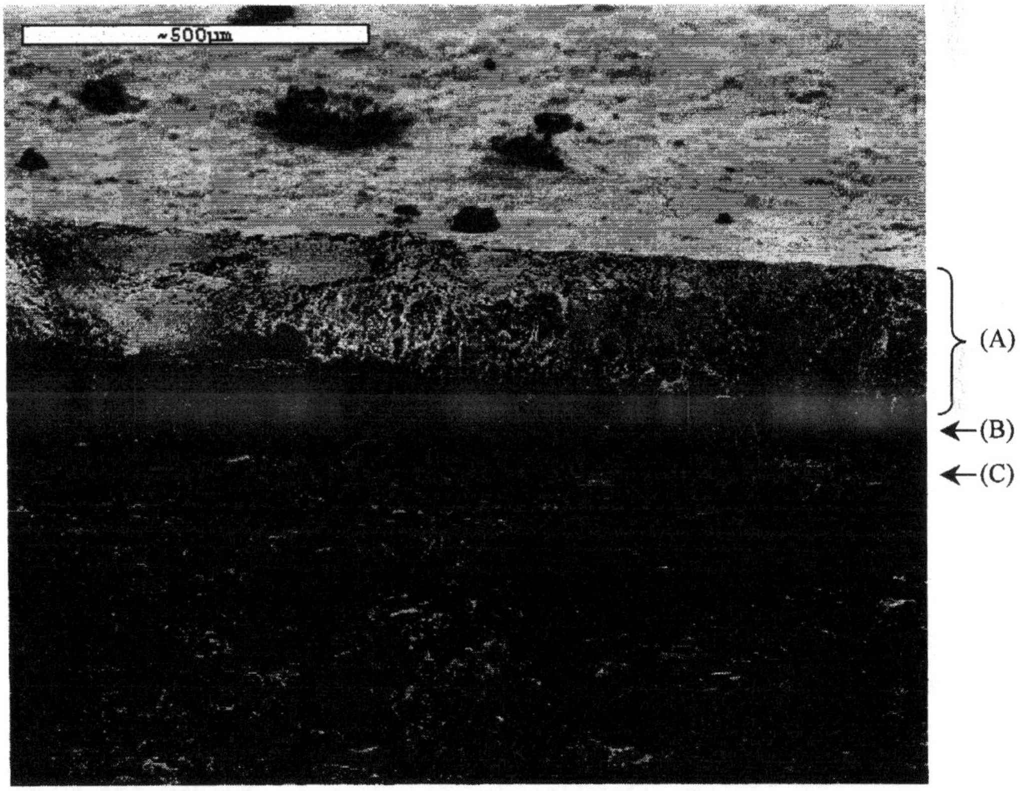

Figure 2. Cross-sectional SEM image of a 19wt%Co/γ-Al_2O_3-coated tubular membrane element. Layer (Thickness): (A) 19%Co/γ-Al_2O_3 catalyst (190 μm); (B) CeraMem membrane (45 μm); and (C) Al_2O_3 membrane (54 μm).

3.2. Catalyst Performance for Fischer-Tropsch Synthesis

Experiments performed with the Pt-Co/γ-Al_2O_3 catalyst reveal that, under identical reaction conditions, the C_2+ yield is higher (Figure 3), and the olefin-to-paraffin ratio is lower (Table 3), when the catalyst is employed in the catalytic membrane reactor. Observed product distributions were reasonably consistent with the Anderson-Schulz-Flory model, from which chain-growth parameters of 0.83-0.85 were calculated.

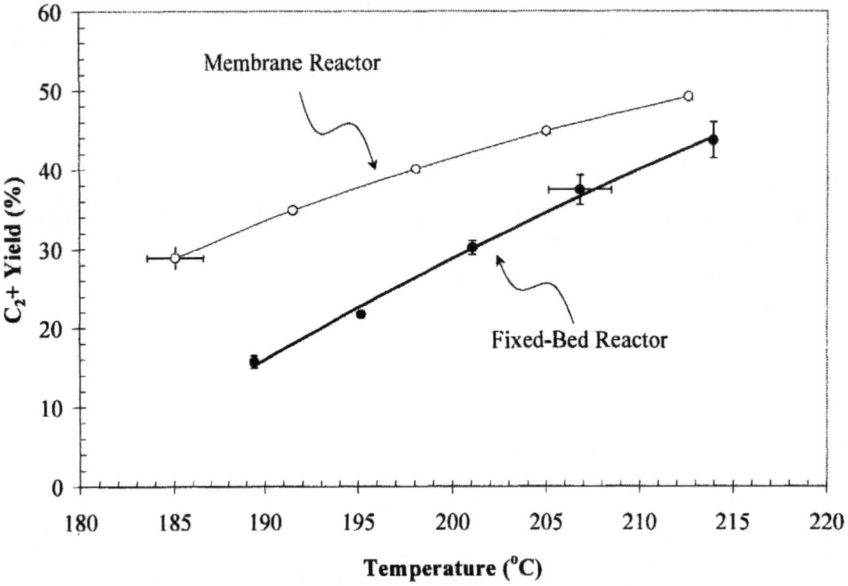

Figure 3. The yield of C_2+, i.e., the moles of CO converted to C_2+ per mole of CO in the feed, over a Pt-Co/γ-Al_2O_3 catalyst in (•) fixed-bed and (○) tubular catalytic membrane reactors. Conditions: $H_2/CO/N_2$ = 1.51/1.00/0.40; P_{inlet} = 321-345 psig; GHSV ≈ 2,446 $cm^3/h \cdot g_{cat}$.

Table 3. Observed olefin-to-paraffin (1-C_nH_{2n} to n-C_nH_{2n+2}) ratios during Fischer-Tropsch synthesis over a Pt-Co/γ-Al_2O_3 catalyst in fixed-bed and tubular catalytic membrane reactors. Conditions: $H_2/CO/N_2$ = 1.51/1.00/0.40; T = 205-207°C; P_{inlet} = 326-341 psig; GHSV ≈ 2,446 $cm^3/h \cdot g_{cat}$.

	1-C_nH_{2n} / n-C_nH_{2n+2}	
n	**Fixed-Bed Reactor**	**Membrane Reactor**
2	0.12 ± 0.01	0.05 ± 0.00
3	1.38 ± 0.18	0.50 ± 0.04
4	0.81 ± 0.04	0.42 ± 0.02
5	0.69 ± 0.04	0.35 ± 0.02
6	0.02 ± 0.00	0.01 ± 0.00

4. SUMMARY

A prototype catalytic membrane reactor for Fischer-Tropsch synthesis has been developed that has a potentially smaller footprint than competing reactor technologies, such that it is a potentially attractive option for the monetization of stranded gas and landfill gas.

ACKNOWLEDGEMENTS

This work was funded by the U.S. Department of Energy under SBIR Grant No. DE-FG02-00ER82971. The authors thank CeraMem Corporation for permission to publish this work; Sasol North America, Inc. for provision of alumina samples; Dr. Burtron H. Davis and his group for helpful discussions and for characterization of the Pt-Co/γ-Al_2O_3 catalyst; Digna X. Fuentes for surface area, pore volume and pore size distribution measurements; and Jayson Zigman for assistance in material characterization via SEM.

REFERENCES

[1] B.H.Davis, "Overview of reactors for Fisher-Tropsch synthesis," *Catal.Today*, **71** (2002) 249.
[2] R.M.de Deugd, F.Kapteijn, and J.A.Moulijn, "Trends in Fischer-Tropsch technology – opportunities for structured reactors," *Topics Catal.*, **26(1-4)** (2003) 29.
[3] For example: *Fossil-Fuel Processing Technical/Professional Services: Comparison of Fischer-Tropsch Reactor Systems. Phase I Final Report.* UOP/SDC. Document DOE/ET/10159-T22 (DE83000822) 1981.
[4] D.J.Wilhelm, D.R.Simbeck, A.D.Karp and R.L.Dickenson, *Fuel Process.Technol.*, **71** (2001) 139.
[5] M.A.Agee, "Converting natural gas into liquid fuels," *PTQ*, **Summer** (1997) 107.
[6] J.Wegrzyn, *Prepr.Pap.-Am.Chem.Soc., Div.Fuel Chem.*, **48(2)** (2003) 917.
[7] P.Stobbe, B.A.Bishop and R.L.Goldsmith, US Patent 6,126,833 (2000); assigned to CeraMem Corporation.
[8] R.L.Espinosa, E. du Toit, J.Santamaria, M.Menendez, J.Coronas and S.Irusta, *Stud.Surf.Sci.Catal.*, **130** (2000) 389.
[9] R.L.Espinosa, J.Santamaria, M.Menendez, J.Coronas and S.Irusta, US 6,403,660 (2002) and WO 99/64380 (1999); assigned to Sasol Technology (Proprietary) Limited.
[10] C.P.Kelkar, J.F.Akyurtlu, A.Akyurtlu and C.E.Hamrin, *Chem.Eng.Comm.* **179** (1992).
[11] M.A.Agee, L.J.Weick and K.L.Agee, US 5,950,732 (1999); assigned to Syntroleum Corporation.
[12] For example, see the article entitled "Alumina Support Found Superior in Cobalt FT Reactor Tests by Energy International," *Gas-to-Liquid News*, January (2000) 12.
[13] G.Jacobs, T.K.Das, Y.Zhang, J.Li, G.Racoillet and B.H.Davis, *Appl.Catal.A:General*, **233** (2002) 263.
[14] J.Zhang, J.Chen, Y.Li and Y.Sun, *J.Nat.Gas Chem.*, **11** (2002) 99.
[15] E.Iglesia, *Appl.Catal.A:General*, **161** (1997) 59.
[16] R.L.Espinoza, J.L.Visagie, P.J.Van Berge, and F.H.Bolder, US Patent 5,733,839 (1998); assigned to Sastech (Proprietary) Limited.

Novel Air/VOC Separation Process

John Bowser, Mark Stutman, and Stuart Nemser*
Compact Membrane Systems, Inc.
325 Water St.
Wilmington, DE USA, 19804
Tel. (302) 999-7996
Fax (302) 999-7955
Email: jbowser@compactmembrane.com
URL: www.compactmembrane.com

Rodger Grantham
Vapor Systems Technologies, Inc.
2030 Webster St.
Dayton, OH USA, 45404
Tel. (937) 228-4673

Abstract

Numerous opportunities exist for separation of air from volatile organic compounds (VOC) in petrochemical handling and chemical processing industries. In this paper we will discuss the application of CMS perfluorinated membranes for the prevention of hydrocarbon emissions in a high volume commercial application.

California Air Resources Board (CARB) has promulgated stringent standards which will go into effect in Spring 2004 for reduction of VOC emissions from underground storage tanks (USTs) at gasoline filling stations. Current gasoline vapor control technologies do not prevent vapor from being released from leaks in hoses and piping and especially from UST vents when USTs pressurize in the normal course of events. Unfortunately, this problem is becoming worse as new automobiles, mandated to have charcoal canister systems on board, are introduced. This is because clean air, rather than saturated vapor, is added to the UST during fueling, resulting in significant pressurization of the UST as liquid gasoline evaporates.

Our approach is to use a CMS membrane in a vapor processor to separate air from the VOCs found in the headspace of the UST. Cleaned air is vented to the atmosphere, while supersaturated gasoline vapors are re-circulated to the UST where they partially condense and remain as saleable product. Removal of the air brings the pressure in the UST and associated vapor piping below atmosphere, thus preventing outward vapor leaks from anywhere in the system.

The first site for certification by CARB for this technology has been established near Sacramento, CA. All equipment is installed and operating at the site, with the certification process having begun in early March 2003. Commercialization is expected to begin in 2004.

Cosinuous Concentration Profile during Chemisorption of Organic Sulfides in Automobile Gasoline onto Palladium due to Diffusion and Convection

Dr. Kal Renganathan Sharma PE
Professor, School of Chemical and Biotechnology
Shanmugha Arts, Science, Technology and Research Academy
SASTRA Deemed University
Tirumalaisamudram, Thanjavur 613402
Tel: 91 04362 264101-107
Fax: 91 04362 264120
Email: jyoti_kalpika@yahoo.com

Abstract

The concentration profile in a Palladium adsorption bed is modeled and the condition where it becomes cosinous at steady state as a function of position is obtained. When Vy < 2 sqrt(Dk'''), and for small Pe,

u = Cos(sqrt(4k* - Pe^2)/2Y)

The solution is bifurcated and Vy needs to be greater than 2 sqrt(Dk''') for the solute concentration to be monotonic with position and not cosinous.

Introduction

The California regulations for sulfur content of automobile gasoline have been further tightened effective 2005. The organic sulfides in the fuel need be removed in a fashion that the emissions should lie within a few parts per billion. The existing technology of catalytic converter and tail pipe emissions need be refined further. In this study a bed of Palladium is used to design a chemisorber to remove the organic sulfides in the automobile gasoline to parts per billion levels. The concentration profile at steady state is derived for the fuel system at steady state and conditions for cosinous behavior are obtained.

The phenomenological relations of Fourier's heat conduction law, Fick's mass diffusion law, and Ohm's electrical conduction law is well established when the cross-effects in the equations between fluxes and forces can be neglected. When a system is close to equilibrium (1) a general theory based upon relations between forces and flows can be formulated. The entropy production per unit volume, σ, can be written as;

$$\sigma = \Sigma F_k J_k \tag{1}$$

in which F_k are forces such as gradient of 1/T and J_k are flows such as heat flow. For a small deviation in the forces from their equilibrium value of zero the flows can be expected to be linear functions of the forces. Hence,

$$J_k = \Sigma L_{kj} F_j \tag{2}$$

Where L_{kj} are constants called phenomenological coefficients. Cross effects such as Seebeck, Peltier, Soret and Dufour may result in one of driving force causing another flow. Onsager (2) provided a well theoretically founded set of reciprocal relations for the phenomenological coefficients. Onsager's theory was based upon the principle of detailed balance or microscopic reversibility and is valid for systems at equilibrium.

During short time scales the system may be far from equilibrium and the high rate unsteady state transient processes cannot be fully described using Fourier's heat conduction law. Fick's law of diffusion is valid only for low rate steady state transfer process. Physically, there exists a time where the linear mass diffusion relationship is not valid. This is because the Fick's law implicitly assumes an infinite speed of propagation of mass in any media. In reality, mass travels at a large but finite velocity. Mathematically this manifests as singularity in concentration for short times and short distances. Maxwell (3) proposed the relaxation concept and used it to predict the Visco elasticity of materials. He assumed that there are no essential differences between the mechanical properties of viscous fluids and solids on the basis of relaxation concept. Relaxation is a progressive reduction of elastic stresses with a constant presented elastic strain. Shear stresses in a viscoelastic body relax over a time interval, τ_r, at a certain finite rate. In an elastic body the shear stresses remain constant, the relaxation time becomes infinite and liquid behavior as an amorphous solid. When relaxation time approaches zero, the body behaves as on ordinary viscous liquid. Cattaneo (4) and Vernotte (5) provided the alternate expression for heat conduction as equation (1) which when combined with the energy balance results in hyperbolic damped wave partial differential equation.

Very little has been done in the inertial effects and not linear relationships between the flux and force of mass diffusion and heat conduction.

Literature Review

A literature review was performed using the Science Citation Index on the topic of hyperbolic equation to represent transient conduction and diffusion. Analytical modeling, computational modeling and experimental studies have been reported. Reviews have been presented by Tzou and Ozisik (6), Joseph and Preziosi (7,8). The problem of the semi-infinite medium subject to a step change in temperature was attempted to be solved by the method of Laplace transforms by Baumeister and Hamill (9,10). Their solution for temperature needs further integration prior to use. Sharma (11) by employing the relativistic coordinate transformation has shown how the hyperbolic damped wave equation can be converted to a Bessel equation in 1 variable. The boundary condition and initial condition can be used to obtain the constants in the solution for temperature and flux at short and long times. Renganathan and Turton (12) found that the thermal inertia effects cannot be ruled out in accounting for the transient heat transfer coefficient in gas solid fluidized beds to immersed surfaces. Sharma (13-31) has shown that for certain values of the dimensions of the finite slab, cylinder and sphere the temperature will undergo pulsations. The damped oscillatory behaviour was also found in simultaneous diffusion and reaction and adsorption. He also presented an infinite order

PDE to capture the transient effects that is glossed over by Fourier's law. The methods of Laplace transforms, method of separation of variables and method of complex temperature were used. The concepts of storage coefficient with units of $W/m^3/K$ in addition to the thermal conductivity W/m/K can be critical parameters in the substrate design of high speed microprocessors especially the portable ones. The work of Taitel (32) and Barletta and Zanchini (33) have mathematical expressions in finite slab transient temperature representation that appear higher than the boundary temperature. In the method of separation of variables, Sharma (22) has shown that the constant has to be set to zero to avoid physically unrealistic values and the result is bifurcated and is monotonic decay in time and for some cases damped oscillatory. He further derived asymptotic solutions in the limits of infinite relaxation time, zero thermal diffusivity. The parabolic solution result at infinite velocity of heat. A serial set of 16 articles at various boundary conditions alongwith the methods used in the solution and the results are available at the Chemistry Preprint Server at **http://www.chemweb.com**. Experimental evidence was shown by Churchill and Brown (34) and Mitra et al (35). Experimental evidence for hyperbolic wave propagative heat equation provided by Mitra et al report relaxation time of the order of 20 – 30 seconds. The work on Kaminski (36) shows that the relaxation time can be as high as 1000 seconds as in falling rate drying (37). The initial estimates of relaxation times by Luikov (38) were in the order of 10^{-12} seconds. The range of relaxation time values is not clear. With the reports of higher relaxation times the use of higher order terms in the extended Fourier's law of heat conduction will be of consequence.

Theory

Applications such as chemisorption of organic sulfides onto a bed of Palladium during automobile fuel desulfurization and during storage of hydrogen as metal hydride involve simultaneous diffusion and convective flow of solute and reaction with the solid adsorbent. The chemisorption rate can be accounted for by a first order reaction rate. Usually the rate is a net removal. Autocatalytic effect if present can be accounted for by letting a net generation and denoting a plus k''' in the governing equation. The constitutive equation used for mass diffusion in the sorption bed corrected for finite speed diffusion can be written as;

$$J = -D\partial y/\partial z \;-\; \tau_r \partial J/\partial t \qquad (3)$$

The mass balance equation may be written as;

(Solute Flow in) - (Solute Flow Out) ± Solute adsorbed = Solute Accumulated (4)

Let,

$u = y/y^*$

where, y^* the equilibrium solute at the entrance of the adsorbent, D the binary diffusivity the governing equation for Autocatalytic chemisorption without any accumulation becomes;

$$d^2 u/dY^2 - Pe\ du/dY + k^*u = 0 \quad (5)$$

The adsorption and desorption cycles will have different signs in the solute adsorbed term.

$Y = y/\mathrm{sqrt}(D\ \tau_r)$; $Pe = Vy/\mathrm{sqrt}(D/tr)$;

$$k^* = k'''tr \quad (6)$$

The solution to the equation is;

$$u = \exp(PeY/2)\ \exp(-\mathrm{sqrt}(Pe^2 - 4k^*)/2\ Y) \quad (7)$$

When $Vy < 2\ \mathrm{sqrt}(Dk''')$, and for small Pe,

$$u = \mathrm{Cos}(\mathrm{sqrt}(4k^* - Pe^2)/2Y) \quad (8)$$

The solution is bifurcated and Vy needs to be greater than 2 sqrt(Dk''') for the solute concentration to be monotonic with position and not cosinous.

References

References

1. D. Kondepudi, I. Prigogine, 1998, Modern Thermodynamics - From Heat Engines to Dissipative Structures, Wiley, USA.

2. Onsager, L., 1931, Phys. Rev. , 37, 405.
3. I.C. Maxwell, A Treatise On Electricity And Magnetism, Oxford University Press, UK, 1873.
4. Cattaneo, C., 1958, Comptes Rendus, 247, 431.
5. P. Vernotte, 1958, Les paradoxes de la theorie continue de l'equation de la chaleur, C. R. Hebd. Seanc. Acad. Sci. Paris, 246, 22, 3154.
6. Tzou D. Y., Ozisik M. N, 1994, ASME Journal of Heat Transfer, 116, 526 – 535.
7. Joseph D. D., Preziosi L., 1989, Heat Waves, Reviews of Modern Physics, 61, 41-73.
8. Joseph D. D., Preziosi L., 1990, Addendum to the Paper on Heat Waves, Reviews of Modern Physics, 62, 375-391.
9. Baumeister K. J, Hamill T. D., 1969, ASME J of Heat Transfer, 91, 543-548.
10. Baumeister K. J, Hamill T. D., 1971, ASME J of Heat Transfer, 93, 126-128.
11. Sharma, K. R., Temperature Solution in Semi-Infinite Medium under CWT using Cattaneo and Vernotte for non-Fourier Heat Conduction, 225th American Chemical Society Meeting, New Orleans, LA, USA, March, 2003, Environmental Division Preprints.
12. Renganathan, K, Turton R, Thermal Inertia Effects In Heat Transfer In Gas Solid Fluidized Beds, Fluidized Bed Combustion Conference, San Francisco, Ca, Usa, 1989
13. Sharma, K. R., Modified Bessel Composite Function Of The First Kind Solution To The Hyperbolic Heat Wave Equation., Abstracts Of Papers Of The American Chemical Society, Bp 17-Fuel, Pg 1, 2001, Vl 222.
14. Sharma, Kr, Quenching Of A Steel Billet, Abstracts Of Papers Of The American Chemical Society,Bp 45-Iec, Pg 1, 2001, Vl 222.
15. Sharma K. R., Critical Radii of Nuclear Fuel Rod Neither Greater than the Shape Limit Nor Less than the Cycling Limit, AIChE Spring Meeting, New Orleans, LA, USA, March/April, 2003.
16. Sharma K. R., Time Taken to Alarm in High Temperature Barrier Coating, AIChE Spring Meeting, New Orleans, LA, USA, March/April, 2003.
17. Sharma K. R, Pulse Injection and Decay in Infinite Medium, AIChE Spring Meeting, New Orleans, LA, USA, March/April, 2003.
18. Sharma K. R., Adsorber Design for Removal of Acrylonitrile from Water, AIChE Spring Meeting, New Orleans, LA, USA, March/April, 2003.
19. Sharma K. R., Storage Coefficient in the Design of 2 GHz Microprocessor, 225th ACS Meeting, New Orleans, LA, USA, March, 2003, PMSE Division Preprints.
20. Sharma K. R., Surface Renewal Theory by Infinite Order PDE to Capture Non-Fickian Diffusion, 225th ACS Meeting, New Orleans, LA, USA, March, 2003.
21. Sharma K.R., Analytical Model for Heat Transfer in CFB to and from a Horizontal Tube, 94th AIChE Annual Meeting, Indianapolis, IN, USA, November 2002.
22. Sharma K. R., Periodicity in Time Domain in Transient Temperature Representation of Finite Slab, 94th AIChE Annual Meeting, Indianapolis, IN, USA, November 2002.
23. Sharma K. R., Heisler Grober Centerline Charts for Transient Temperature Representation Corrected for Finite Speed of Heat Propagation, 94th AIChE Annual Meeting, Indianapolis, IN, USA, November 2002.

24. Sharma K. R., Concentration Polarization Layer in Reverse Osmosis, 94th AIChE Annual Meeting, Indianapolis, IN, USA, November 2002.
25. Sharma K.R., CWF Problem in Semi-infinite Medium using Hyperbolic Heat Wave Propagative Equation, 223rd ACS Meeting, Orlando, FL, USA, April 2002.
26. Sharma K. R., Thermal Time Constant and Relaxation Time of a Sphere, 34th Middle Atlantic Regional Meeting of ACS, Baltimore, MD, USA, June 2001.
27. Sharma, Kr, Hyperbolic Wave Propagative Partial Differential Equation Solution In Elution Chromatography, Abstracts Of Papers Of The American Chemical Society, Bp 79-Anyl, Pg 1, Py 2001, Vl221.
28. Sharma, Kr, Turton, R, Mesoscopic Approach To Correlate Surface Heat Transfer Coefficients With Pressure Fluctuations In Dense Gas-Solid Fluidized Beds, Powder Technology, Pg 10, 1998, Pd Sep 15, Vl 99, Is 2.
29. Sharma, Kr, Relative Contributions From Particle Conduction And Gas Convection To The Heat Transfer Coefficient Between Dense Gas- Solid Fluidized Beds And Surfaces, Powder Technology, Pg 6, Powder Technol, Py 1997, Vl 91,Is 1.
30. Sharma K.R., Calculation of Relaxation Time and Thermal Diffusivity, 93rd AIChE Annual Meeting, Reno, NV, USA, November, 2001.
31. Sharma K. R. 2003, Finite Speed Heat Conduction – XVI Convective Boundary Condition and Conditions for Pulsations in Temperature.
32. Taitel Y., 1972, International Journal of Heat and Mass Transfer, 15, 369-371.
33. Barletta A., Zanchini A., 1997, International Journal of Heat and Mass Transfer, 40, 5, 1007-1016.
34. Brown, Churchill, Experimental Evidence of Second Sound, AIChE Annual Meeting, San Francisco, CA, USA, November, 1989.
35. Mitra K., Kumar S., Vedavarz A., Moallemi M. K., 1995, ASME Journal of Heat Transfer, 117, 568 – 573.
36. Kaminski W., 1990, ASME J of Heat Transfer, 112, 555-560.
37. Mitura E., Michalowski S., Kaminski W, 1988, Drying Technology, 6, 1, 113-137.
38. Luikov A.V., 1966, International Journal of Heat and Mass Transfer, 9, 139-152.

Acknowledgements

Acknowledgements are extended theVice Chancellor of SASTRA Deemed University, R. Sethuraman for his never failing encouragement and the blessings of their Holiness Jayendra Saraswathi Swamigal and Sankara Vijayendra Saraswathi Swamigal of the Kancheepuram Sankara Matt.

Environmentally Friendly Methods for Large Scale tert-Butoxycarbonyl (BOC) Deprotections

Eric Dias, Kevin Heffenbach and David Am Ende*
Pfizer, Inc.
Eastern Point Road
Groton, CT USA, 06340
Tel. (860) 715-3583
Email: eric_l_dias@groton.pfizer.com

Abstract

Isobutylene off-gas amounts liberated during the methanesulfonic acid-catalyzed deprotection of N-BOC-pyrrolidine in several common organic solvents were quantified using on-line gas phase mass spectroscopy. These results provided a basic strategy for controlling isobutylene emissions in situ, and led to the development of an effective scrubber that was employed on the Pilot Plant scale.

A Novel Process of Enzymatic Synthesis Coupled with on-Line Separation for Producing N-lauroyl-β-amino Propionitrile through a Packed Bed reactor

Yun FANG[a], Yong-mei. XIA[a,b], Zhi-guo[a] XU, Xue-yi[a] HU, Yong-an[a] SHEN

a School of Chemical and Material Engineering, Southern Yangtze University, 170# Huihe Road, Wuxi 214036, China

b Department of Chemical Engineering, University of Michigan, 3085 H.H. Dow, Ann Arbor, MI 48109, USA

Abstract: *N*-lauroyl-β-amino propionitrile is the intermediate to synthesize sodium N-lauroyl-β-alanine, an antimicrobial surfactant. We describe a novel process of enzymatic synthesis coupled with on-line separation to produce N-lauroyl-β-amino propionitrile through a cascade connection of an enzyme packed bed reactor (EPBR) with a crystal separator. The substrate solution was feed into the inlet of the reactor and the crystal product with high purity was obtained from outlet of the separator with a yield of 91.7% under the optimum conditions. The immobilized lipase can be utilized repeatedly and the solvent carrying the unreacted substrates was recovered and reused on-line.

Keywords: Enzymatic synthesis, *N*-Lauroyl-β-amino propionitrile, Lipase, β-amino propionitrile, organic media

1. Introduction

N-lauroyl-β-amino propionitrile is the precursor of N-lauroyl-β-alanine, the latter can be used as an antimicrobial surfactant in many fields such as food, cosmetics, pharmacy and textile after saponification[1-5].

Taeko Izumi et al[6] used lipase from *Candida antarctica* (CAL) as the catalyst for amidation of β-amino propionitrile to synthesize *N*-lauroyl-β-amino propionitrile. They claimed that room temperature was a suitable reaction temperature, and by using equimolar reactants in diisopropyl ether along with 7000 U of the lipase per gram of β-amino propionitrile in a flask, with equal weight of deionized water sprayed on the enzyme, the best yield (99.3%, determined by TLC) was achieved after 24 h. Actually according to our result [7], at room temperature the reaction rate was quite slow even with such a high amount of lipase, meanwhile the product would unexpectedly precipitated and mixed with the immobilized lipase so that a great loss of lipase activity could not be avoided when separating the product from the catalyst. We optimized the reaction parameters with the same reaction system in a batch reactor[7], with a quite lower lipase dosage (200 U/g β-amino propionitrile), equimolar reactants in diisopropyl ether (35 g β-amino propionitrile/L), the highest conversion (96.8%) of β-amino propionitrile was obtained after 24 h at 70℃.

As mentioned above, it is unavoidable that at lower temperature the product *N*-lauroyl-β-amino propionitrile would precipitated from the reaction solution in batch reaction. Not only the mechanical stir but also the separation process, therefore, caused the enzyme activity loss. Thus if applied the enzymatic amidation at a relative high temperature in an EPBR, modified the operation parameters to avoid precipitation in the EPBR and then separated the crystal product at a lower temperature in a separator, we can expect to get high yield of the product with desired purity as well as the least lipase loss . Hence a EPBR filled with CAL will be used in this paper to carry out this reaction together with an on-line crystal separator to separate product, consequently the lipase can be utilized repeatedly and the solvent carrying the unreacted substrates can be recovered and reused on-line after some simple processes.

2. Experimental

2.1. Chemicals

Immobilized lipase from *Candida Antarctica* (CAL) with olive oil hydrolysis activity 7000 U/g was a gift from Novo Nordisk Co. β-amino propionitrile was synthesized in this lab and distilled before use. Methyl laurate was purchased from Haiyan Fine Chemicals Co, Zhejiang, China; the purity was more than 99.1% (GC) after rectification . All other reagents are of analytical grade and were purchased from Shanghai Chemicals Co., China. Reactants unless otherwise stated were dehydrated with 4A zeolite before use.

2.2. Analysis

Water content was determined by Karl Fischer titration.

Conversion of β-amino propionitrile was determined by HCl (in isopropyl alcohol) titration with following equation:

$$X = 100(1 - A_1/A_0)$$

Here A_1 was the consumed HCl (mmol) per gram of the product mixture obtained from the outlet of the reactor; A_0 was the consumed HCl (mmol) per gram of the reactant mixture before reaction.

2.3. Enzymatic synthesis of N-lauroyl- β -amino propionitrile

The reaction system was consisted of an elevated tank① (for mixing and storing raw reactants, equipped with a refluxing tube and a mechanical stir), an EPBR③ （0.7 cm in diameter and 5~8cm in height, filled with CAL）, a crystallization tank④, a crystal separator⑥ and a methanol absorption column⑦ filled with $CaCl_2$ as showed in Fig.1. The mixture of equimolar of methyl laurate⑧and β-amino propionitrile in diisopropyl ether⑨ （400mL/g β-amino propionitrile） were well mixed in the elevated tank and warmed-up to 65℃. The reactant mixture was then delivered through the EPBR (65℃) with a given flow rate② to the crystallization tank where the crystal *N*-lauroyl- β-amino propionitrile precipitated at 25±1℃. The slurry was pumped to the crystal separator by a slurry pump⑤ and separated by the filter film in the separator. The precipitate⑩ was washed with cold 95% of ethanol and dried for succedent hydrolysis to produce sodium N-lauroyl- β-alanine， while the filtrate was delivered to an adsorption column to remove the byproduct methanol. After determining the content of β-amino propionitrile, the mixture came from the outlet of the adsorption column was made up a loss with fresh reactants and solvent before ready for reaction again. The temperature at all parts in the system excluded the section from elevated tank to the outlet of EPBR was maintained at 25±1℃。

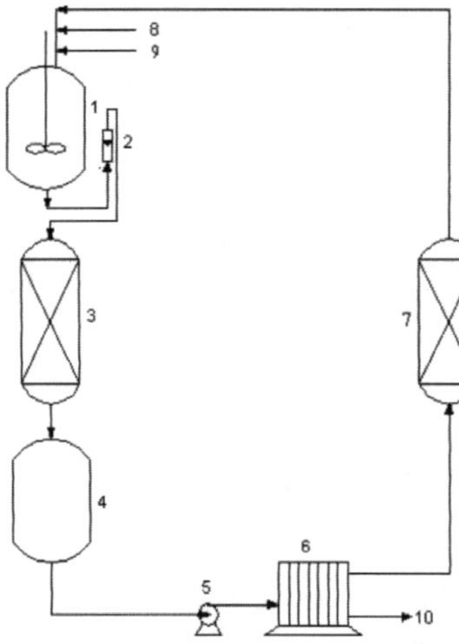

Fig. 1. The scheme of the reaction system

1. elevated tank 2. flow meter and valve 3. EPBR 4. crystallization tank 5. slurry pump 6. crystal separator 7. absorption column 8. fresh methyl laurate 9. fresh β-amino propionitrile and solvent 10. crystal product

3. Results and discussion

3.1. Flow style and substrate concentration

The reactant flow came from the elevated tank to the EPBR and the flow rate was monitored by a flow meter. The concentration of the product increased along the reactor vertically through the inlet to outlet. To avoid precipitation in the reactor, it is required a suitable initial substrates concentration and a suitable total yield when the reaction mixture reached the reactor outlet since high product content in the reaction mixture may cause precipitation.

A downstream flow was applied to avoid back-mixing. The flow rate will be discussed in this paper later. As to the substrate concentration, although 35 g β-amino propionitrile per litter diisopropyl ether was the highest concentration which induced the least by-product in a batch reactor [7], considering the product solubility (as shown in Fig. 2) under the suitable reaction temperature as well as with the operation flexibility, a diluted substrate concentration was adopted as 2.5g β-amino propionitrile per litter diisopropyl ether based on the results from several testing (data unshown).

3.2 Reaction temperature

The optimal reaction temperature in batch reactor was 70℃ [7] as showed in Fig. 2. But in column reactor, the diisopropyl ether would partly vaporized at 70℃ during the delivery and reaction so that it would induce bubbles in the column. Along with the considering of the solubility of *N*-lauroyl- β-amino propionitrile and reaction rate, as well as the enzyme thermostability (data unshown), the reaction and delivery (from elevated tank to the reactor) temperature was set up at 65 ℃. At this point, neither solvent vaporization nor product precipitation was observed during the reaction.

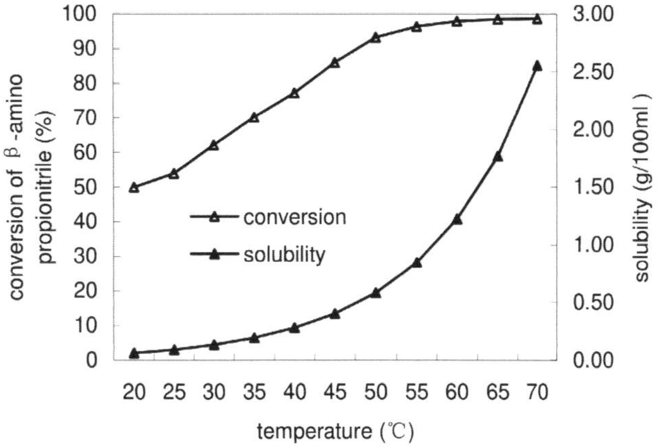

Fig. 2. The solubility of *N*-lauroyl- β-amino propionitrile in diisopropyl ether and the effect of temperature on conversion of β-amino propionitrile in batch reactor.

Reaction conditions in tank reactor: 5mmol of β-amino propionitrile (0.35g) and methyl laurate (1.07 g), 10ml of diisopropyl ether, 0.01g of CAL (200u/g β-amino propionitrile), 0.021g of water , 24h.

3.3 water dosage

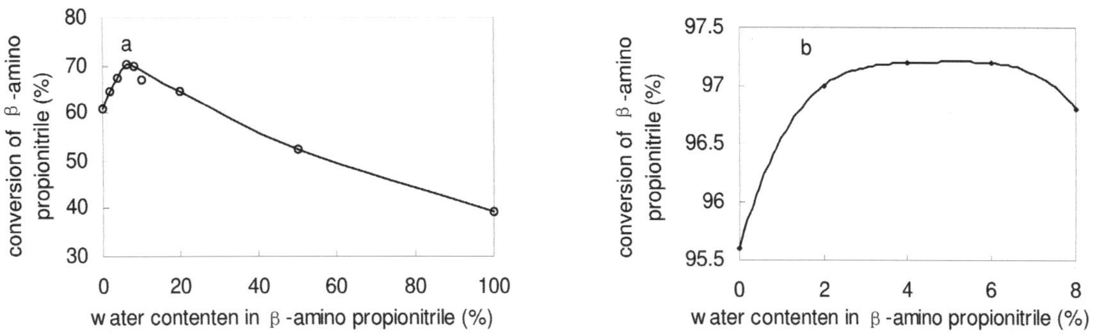

Fig. 3. The effect of water content on the conversion of β-amino propionitrile in tank reactor.

Reaction conditions in batch reactor: 5mmol of β-amino propionitrile (0.35g) and methyl laurate (1.07g), 10ml of diisopropyl ether, a: 0.15g of CAL (3000u/g β-amino propionitrile), 35℃, 6h. b: 0.01g of CAL (200u/g β-amino propionitrile), 70℃,

24h.

Water is essential to enzyme catalytic reactions. CAL is usually used together with equal weight of deionized water according to the suggestion from Novo Nordisk Co[8], so did by some investigators in their research about CAL generated amidation of β-amino propionitrile[5]. This means that 0.4g of CAL associated with 0.4g of water was added in a system containing 0.4g of β-amino propionitrile and 20ml of diisopropyl ether[5]. Whereas enzymatic reactions performed in nonpolar solvent system usually favor lower initial water content in the reaction mixture[9]. Meanwhile as indicated in Fig3, about 6% of exogenous water in β-amino propionitrile (0.06g of water to 1g of β-amino propionitrile and 0.15g of CAL) was optimal to the reaction in batch reactor. From Fig. 3b, without exogenous water, the conversion (95.6%)of β-amino propionitrile was not much lower than that at the optimal water dosage (6% of water in β-amino propionitrile, 97.2%), therefore no water was added to generate this reaction in the column reactor. The reason that without exogenous water in β-amino propionitrile also yielded a high conversion is because the commercial CAL was not a "dry" lipase but contained 2% of water(w/w). Additionally, the water activity of the reaction system finally was equal to water activity in atmosphere since the system had been contacted with the atmospheric environment through the venting tube (refluxing tube) of the elevated tank.

3.4 The effect of Packed bed height on total conversion of β-amino propionitrile

As mentioned in section 3.1, flow rate is very important for controlling a sufficiently high but not the highest total yield. As a matter of fact, another operation parameter, the reactor bed height has to be considered along with the flow rate.

Table 1 showed effect of reactor bed height on the conversion of β-amino propionitrile at the reactor outlet under different bed height.

Table 1. The effect of reactor bed height on total conversion of β-amino propionitrile

No	Reactor bed height (cm)	Flow rate of reactant mixture (ml/min)	Conversion (%)
1	3.5	2.7	68.6
2		2.3	78.7
3		1.9	89.9
4	4.5	2.4	81.6
5		1.9	90.7
6		1.7	91.7
7		1.0	94.3
8	6.0	2.3	91.4
9		1.9	92.7
10		1.5	94.0

Packed reactor, equimolar of reactants, 65℃, β-amino propionitrile: diisopropyl ether=1：400(w：v).

From Table 1, No.6 was finally chosen as the suitable operation condition with considering of reasonable higher conversion, lower enzyme amount and shorter operation period. Thus the conversion of β-amino propionitrile at the outlet of the reactor was 91.7% at 1.7 ml/min of flow rate while the reactor bed height was 4.5 cm. The raw crystal product was then washed with cold (5-10 ℃) 95% ethanol, no trace of β-amino propionitrile or methyl laurate was found after drying the crystal under vacuum. And no obvious loss of the product was observed during the washing and drying.

3.5 The operation stability of the system

It was observed that recycle the filtrate from the crystal separator without removing methanol (a byproduct in this reaction) induced some decrease of the conversion of β-amino propionitrile because that methanol affected the lipase activity. Thus the filtrate was processed through an adsorption column filled with $CaCl_2$ to remove methanol, then mixed with fresh reactants to the required substrate concentration before feeding to the elevated tank. No obvious ill effect of the reusing of reactants on the conversion was observed after applying the reaction in a batch reactor with this mixture and fresh lipase (data unshown). Therefore an adsorption column was added between the elevated tank and the crystal separator. After running the reaction system 50h, the conversion of β-amino propionitrile did not drop dramatically as illustrated in Fig. 4.

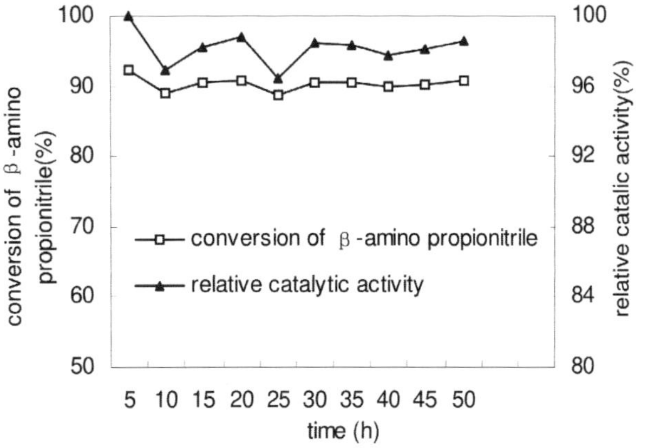

Fig. 4. The conversion of β-amino propionitrile versus operation time

Packed reactor, equimolar of reactants, 65℃, β-amino propionitrile: diisopropyl ether=1 : 400(w : v).

4. Conclusion

CAL catalyzed amidation of β-amino propionitrile to N-lauroyl-β-amino propionitrile in a packed bed reactor coupled with on-line separation system was investigated. The conversion of β-amino propionitrile at the outlet of the reactor reached 91.7% with following operation parameters: 1.7 ml/min of flow rate, 4.5 cm of reactor bed height (reactor: 0.7 cm in diameter and 6.5 cm in height), reaction temperature was 65℃, crystallization temperature was about 25℃.

The conversion of β-amino propionitrile did not drop dramatically after running the reaction system 50h. The final crystal product can be simply purified with ethanol washing and drying.

References

[1] M.S. Jarrell, B.C. Gates, E.D. Nicholson, J. Am. Chem. Soc. 100 (1978) 5727.
[2] Iwata Masaaki, Muramatsu Kazuhiro, Agata Kazuko , Jpn Appl. 86/257,987 29 Oct. 1986.
[3] Wu Whei Mei,Murakami Teruo,Yamajo Ryoko , J. Pham. Sci. 78(1989) 499.
[4] Beger Joerg, Noumann Kennnate , Tenside. 23 (1986) 135.
[5] Taniguchi Hiroaki, Maruyama Teruhito, Jpn appl. 810 25 Apr 1991.
[6] Taeko Izumi, Yoshihiro Yagimuma, and Masanori Hagal.,J.Am.Oil Chem. Soc.74(1997)875
[7] Zhi-guo Xu, Master Thesis, Wuxi University of Light Industry, 2002.
[8] Novo Nodisk product sheet.
[9] Y.M. Xia, Ph.D. Thesis, Wuxi University of Light Industry, 2000.